D1743359

Drummer Boy illustration by Thomas Nast

THE REAL BOYS OF THE
CIVIL WAR

RESEARCHED AND EDITED BY

J. ARTHUR MOORE
author and educator

ISBN: 978-1-952874-90-1 (paperback - bnw)
978-1-952874-91-8 (hardback - bnw)
978-1-952874-84-0 (paperback - colored)
978-1-952874-85-7 (hardback - colored)
978-1-952874-86-4 (eBook)

This is a work of non-fiction historical research.

Cover design by Gian Carlo Tan

Published by:

OMNIBOOK Co.
99 Wall Street, Suite 118
New York, NY 10005 USA
+1-866-216-9965
www.omnibook.org

For e-book purchase: Kindle on Amazon, Barnes and Noble
Book purchase: http://www.jarthurmoore.com
www.Amazon.com and www.Barnes&Noble.com

Omnibook titles may be purchased in bulk for educational, business, fund-raising, or sales promotional use.
For more information please e-mail admin@omnibook.org

BACKGROUND NOTE

This book has evolved over several years of research and collecting of materials. The opening research evolved from a graduate study research paper on the historiography of the boys of the American Civil War – Department of History, West Chester University, West Chester, Pennsylvania, course number His 500-80, December 10, 2014. It was eventually published on line as part of the Essential Civil War Curriculum, a project of Virginia Center for Civil War Studies at Virginia Tech. This site contains over three hundred topics for study. https://www.essentialcivilwarcurriculum.com/.

The research is followed by a source document of several pages, divided into primary and secondary and online resources. There are many excellent items for anyone interested in learning more about these boys and their lives, written both by them and about them. One researcher, Professor Jay S. Hoar [May 15, 1933-April 12, 2023], made it his life's work to research and publish the stories of the youngest to serve and the last to pass on, with a multitude of works, culminating in 2010 with his 2000 page trilogy, *Sunset and Dusk of the Blue and the Gray*. One group of boys extensively covered in his research but ignored by most, was the Home Guard which included boys mostly age 6 to 11 who did guard duty, courier service, and teamster service driving wagons around the countryside to gather supplies for their respective armies. The author had the honor of years of friendship and phone conversations as well as the exchange of their published works.

Following the research is a section sharing the aspects of the war from the boys' point of view. Focus sections will look at those who were or believed to be the youngest in several categories, recipients of the Medal of Honor, several who lost their lives, and a general collection of several stories. Finally, one of the Home Guard boys from Professor Hoar's research is shared followed by a photo gallery of many, both known and unknown.

DEDICATION

The Real Boys of the Civil War is dedicated in His love and in friendship to the hundreds of thousands of boys who served in the conflict, to those who have sought to preserve their history through their research and publications, and to all who seek to learn of their existence and their lives that they not be permanently gone forever.

TABLE OF CONTENTS

Drum Corps of 61st New York Infantry, Falmouth, Va., March, 1863

BOYS AT WAR
Civil War Historiography Evolves about Boys of the American Civil War

Prepared by Joel A. Moore
HIS 500-80

Research Paper
Department of History, West Chester University
West Chester, Pennsylvania
December 10, 2014

Boys of the Civil War

By Joel A. Moore

Willie Lawn, 10
Wounded and lost part of his
right arm near Suffolk Virginia

David Wood, 10
Aside to his father, Colonel 6th
Missouri cavalry, created a sutler
business and earned $2000

Johnny Clem, 12
A runaway with 22nd Michigan,
shot an enemy colonel, sergeant
with General Thomas' staff

Willie Johnston, 12
Drummer 3rd Vermont
awarded the Medal of Honor
Peninsula Campaign

Washington H. Potts, 14
Highly talented
drummer and bugler with
Independent Battery I

Elisha Stockwell, 15
Soldier from Alma
Wisconsin, a runaway, lied
to enlist and kept a diary

Rashio Crane, 15
Company D 7th Wisconsin,
died in Andersonville Prison

Charles Bardeen, 14
Fifer/Drummer 1st
Masachusetts, published
A Little Fifer's War Diary

Henry Messhage, 12
1st Class Boy and powder monkey
carried munitions on ships

Edwin Francis Jennison, 15
2nd Louisiana Infantry killed
at The Battle of Malvern Hill

Tad Lincoln, 12
Commissioned 2nd Lieutenant,
son of President Lincoln

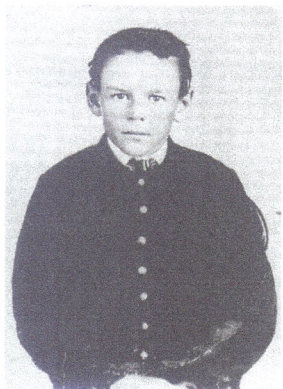

Benjamin Knox, 15
Company H 20th Ohio killed
in the trenches at Atlanta

James Marten in his collection of essays about *Children and Youth During the Civil War Era* wrote that the question has been raised, "Have social historians lost the Civil War?" Civil War historians and historians of children and youth have generally ignored the children and youth who participated in the war. Some have begun to explore the topic in recent years. Emmy Werner shared the insights of Civil War children, particularly the combatants, in *Reluctant Witnesses*. Historians Jay S. Hoar, Dennis M. Keesee, and G. Clifton Wisler have explored the experiences of young participants within the armies of the war. The histories of these children are placed within the context of the conflict and the social environment of the times. Jim Murphy addresses the conditions faced by these young combatants from the event through which they became soldiers to the preparation for combat to their entry into battle. Bell Irvin Wiley shares his in-depth research as to who these individuals were, where they came from, and the attitudes and backgrounds they brought with them. Within their works these historians explain how these boys entered the war, the evolution of attitudes and opinions both external and internal, and impact and conditions of war. Essays from Marten's work focus on the schools of thought of modern childhood, the impact of death on children and society, and the *Companion*, a periodical for children, and their impact on the children of the war and the war's impact on the times. Finally, within this narrative the shifting perspectives of the boys themselves reflect, from their own writings, the impact of the events of their historical time and place.

The American Civil War has sometimes been called the Boys' War because so many boys ages seventeen and under saw active service in the two armies. Many historians have estimated their number to be 250,000 or more.[1] How could this happen? How were they recruited? At the time of the war, the presence of young boys was an accepted practice.[2]

Dennis M. Keesee explained the rules at the time. Regulations over the years required soldiers to be at least eighteen years old, no younger than seventeen with written parental permission, pass a physical examination, and be at least five foot three inches tall. Musicians could be as young as twelve, were to pass a physical examination, but were excused from the height requirement.[3] Between the two armies, there was need for 60,000 musicians.[4]

Keesee went on to describe how recruitment worked. Companies and regiments were usually recruited from local communities by local businessmen, politicians, teachers, and ministers. Known and respected within their communities, they could easily gather enough recruits to fill the ranks and on that basis, were issued commissions as the commanding officers of the units they recruited. One of many ways in which boys entered the army was with their teachers or ministers. [5] Hoar added that many young teens were accepted by means of their own lies, by recruiters who looked the other way, by accompanying other family members, or by presenting themselves as orphans. The very young tended to be valets or servant to commissioned fathers, mascots, some musicians, and some who took the names of older family members or friends.[6] G. Clifton Wisler added that some enlisted in another community where they were not known and less likely to be questioned. And some declined pay and did not sign muster books so there was no record to check.[7] Finally, an unusual group of unassigned youth within the Confederate army were the drillmasters borrowed temporarily from Southern military schools to assist in the training of new recruits. Cadets thirteen years old and up were fully respected by older men as they drilled and trained them upon their entrance into the army. Some managed to join the units they trained and follow them into active service.[8]

There were other opinions and attitudes to overcome. In 1861, President Lincoln forbade the enlistment of soldiers under eighteen without written consent and a year later under any circumstances. President Davis, in one of his earliest speeches, argued against recruiting any under the age of eighteen.[9] Neither one had any influence over the entrance of young boys into the war. One group did have some effect.

Fifteen-year-old Elisha Stockwell explained in his memoirs that when he signed up, his father was there and crossed his name out.[10] Five months later, he ran away with a friend whose captain got him in by lying a little. Nine-year-old Johnny Clem ran away from home and was able to join up by starting out as a camp helper until he was older and mustered in as a drummer.[11] Ten-year-old David Wood was told in no uncertain terms by his father, a regimental cavalry commander, that he could not go with the regiment. He sneaked into the back of the column on a long march, was discovered by his father, then enlisted as an orderly when his father realized he would not stay at home.[12] Seventeen-year-old Aaron Stauffer was not only prohibited from joining the war by his mother, but also by his Mennonite church. He ran away and joined anyway, lying about his age.[13] Twelve-year-old Gustave Schurmann was an exceptionally talented musician. His father had taught him. His father had enlisted for the war, but his mother forbade his going. Finally, his parents decided that he might run away anyway and join up with strangers and decided they would rather he go with his father and be with family.[14] When Mancil Root, not yet nine, volunteered to help fill the quota for his county, everyone cheered, but his recently widowed mother objected. When she was assured that his unit would never leave the state and she would receive his $500.00 enlistment bonus, she consented.[15] Parent opinion was a major factor in boys joining the war. Some objected, some supported, some compromised, some were persuaded. In most cases, it didn't matter. The boy went anyway.

North or South, these boys were looking for adventure and a break from the boring life of farm or school.[16] A few, like fourteen-year-old John Wise, had mixed emotions. Though excited about war, as he watched the national flag lowered from the custom house in his home town and the stars and bars raised in its place, he suddenly realized that friends in the nearby Gasport Navy Yard who had always been welcome in their house, had suddenly become the enemy.[17]

For some, as described in the diary of fifteen-year-old William Bircher, the war experience began with weeks of drill, camp, marching and camping in fair weather and torrential rain, suffering boredom with sickness and poor food, and involvement in raids and small fights. The first encounter in battle was the arrival just after the battle, in time to see the wreckage and carnage that littered the field. For the first time the boy's excitement for adventure, having been passed to boredom, was changed to retched sickness and the sudden reality of the dreadful destruction, suffering, sorrow, and tears that were the real face of war.[18] For another, described in the memoir of fourteen-year-old Johnnie Wickersham, war exploded around him before his unit had left home. He described his first battle, "having come on like a flash of lightning," wherein he shot his first enemy and was so excited as to be oblivious to his surroundings.[19] Additionally, a number of boys, found themselves homesick, unable to cope with the rigors of military life, sick from related health problems, or simply unfit and incapable of carrying out the duties for which they had volunteered.[20] Many either had a change of heart or were too sick and were mustered out.[21] Others, like John Delhaney and Elisha Stockwell noted their feelings in their journals and soldiered on.[22] Boys attitudes changed dramatically from their initial sense and want of adventure, excitement, and glory as the reality of war set in.

Attitudes affecting policies during the war changed many times. One area had to do with the treatment of prisoners and within this, the effect young prisoners had on prison commanders. Early in the war, prisoners were exchanged shortly after a given battle, but Grant had stopped the practice in the summer of 1864.[23] Afterwards prisoners were sent to prison camps. Conditions were horrific. In May of 1864, Fifteen-year-old Rashio Crane was captured while trying to help a wounded comrade off the battlefield. Sent to Andersonville Prison, he died two months later from intense suffering and deprivation.[24] Eleven-year-old Brice E. Davis was captured at Shiloh in 1862 and sent to Libby Prison. Because of his small size and young age, he was allowed to play and run around outside the prison walls. He was exchanged a few months later.[25] In the fall of 1863, seventeen-year-old Michael Dougherty was captured and ended up by late spring at Andersonville Prison. He received no special treatment, but through his own determination managed to survive until released April 12, 1865.[26] Fourteen-year-old Ransom T. Powell was captured in the winter of 1864. He was taken to Belle Isle Prison where he and other young boys were treated very differently from the older prisoners. Given the privilege of running around the headquarters enclosure, they had free reign of the bake house and the cook house where they got plenty to eat and good times including sailing on the James River in small skiffs. Weeks later he was among the first groups to be shipped off to Andersonville Prison, where once again, he benefited from a different attitude toward younger prisoners, finding himself living outside the prison in the general's quarters serving as a house boy doing odd jobs.[27]

Who were these boys? In a war often referred to as brother against brother and American against American, that wasn't truly the case. A glance through the anthologies chronicling so many of these boys, it becomes obvious that many were foreign born. Wiley presents an in-depth study of the nature of the southern soldier versus that of the northern soldier in his companion books, *The Life of Johnny Reb* and *The Life of Billy Yank*.

In researching the manner of men who fought for the Confederacy, Wiley found that thousands were actually born in the northern states. The ages ranged from twelve to seventy-three. The boys in the ranks served well with many examples of significant bravery. A few left early when they found the reality of war was far from the adventure expected. Significant in the South was the number of military academies. Later in the war, their cadet corps were established as reserve units and some were actually called into service. Social class had a significant effect in the Confederacy. Plantation born soldiers were used to giving orders, not taking them. Many were unruly and when commanded by officers not of the upper class, were outright rebellious.[28] Young boys and youth were not always the most disciplined soldiers. They had romantic expectations of the war and weren't accustomed to taking orders from superiors. While they wanted to fight, they wanted to fight on their own terms.[29]

Wiley wrote of the great diversity of the Union army in nationality, race, creed, occupation, dress, habits, temperament, education, wealth, and social status. Most of the youngest boys were musicians, but they supplemented their position in a variety of ways—with barbering, carrying water for soldiers, sharpening surgical instruments, helping the wounded, burying the dead, drawing maps, selling food delicacies, and gambling. They were the pets and favorites of their units, some slept in the tents of their officers, and on a long march, when tired, rode a horse provided by an officer. As with the southern boys, a few left after their first experiences in battle, but most acquitted themselves very well. Many took up a rifle in combat, gallant and effective on the front line. Some took their place with the artillery. Some were recognized with the Medal of Honor. Some rose through the ranks as officers becoming colonels or lieutenant generals by age eighteen. And some gave their lives in the service of their country.[30]

The Civil War was unlike any previous war. It was the largest and bloodiest conflict ever fought on American soil with millions of soldiers and hundreds of thousands of boys within the two armies.[31] Emmy Werner in her work *Reluctant Witnesses* explains why there exists such a wealth of information left behind by the boys who fought in the war as well as other children affected by the war. The North had one of the highest literacy rates in the world by way of compulsory education and the regular attendance of children in school. The writings from the South were predominately from children of privileged families who sent their boys to private academies or educated them at home by way of private tutors. Late in the war the writings of emancipated slave children and contraband soldiers were made possible in narratives recorded by the Federal Writer's Project.[32]

At the time of the war, little attention had been given to children and youth— their insights into the war and their place in social history.[33] Emmy Werner in her work *Reluctant Witnesses* chronicled the insights of Civil War children, especially the combatants. Her primary resources were diaries, journals, letters, and reminiscences of the children and, when relevant, eyewitness accounts of family members.[34] An example given was from the siege of Vicksburg, from twelve-year-old Fred Grant.

Suddenly a small boy, no larger than himself, came running from the front, the blood streaming from a wound in his left side, crying: "General, our regiment is out of ammunition." Noted Fred: "The little fellow, becoming weak from the loss of blood looked up and said, "Caliber 68," and as he tottered he was seized by two soldiers and carried to the rear. I went up to my father…and found to my surprise that his eyes were suffused with tears of sympathy for the brave boy."[35]

Another was a series of examples to share the humanitarian attitude of boys during the war. At Gettysburg, hungry Confederate soldiers shared candies with children and helped their families to safety beyond the battle line. Under fire at Sharpsburg, fifteen-year-old Union soldier Thomas Galway shared the water from his canteen with a wounded southerner. At Mayre's Heights at Fredericksburg, teenage Confederate soldier Richard Kirkland stepped onto the open battlefield with a collection of canteens to share water with the wounded and dying Union soldiers on the field.[36] In a letter home, Union teenager James Newton described how Confederate and Union soldiers would gather together under a flag of truce for a long talk and would agree among themselves that if the war were left to the enlisted men, they would soon go home.[37]

*   *   *

Most who read this have no idea that these boys served. For those who seek more knowledge, their history is out there, it begins with the boys themselves.

From 1861 to 1865 the historians of the Civil War were the participants, soldier or civilian. Their letters and journals, the photographs, sketches, newspaper articles and more form the primary sources of the era. In a war in which more than 200,000 boys age seventeen and under served in the opposing armies, their stories are a significant part of that history.

The history of the battlefields and military life of the Civil War is written in the journals and letters of Charles William Bardeen, William Bircher, Elisha Stockwell, Johnnie Wickersham, and others. Michael Dougherty recorded life in Andersonville Prison. Johnny Clem and Robert Hendershot became a part of history by deeds that were captured in newspaper articles. Newspapers also reported the sad fate of others as in the cases of Charlie King and Clarence McKenzie. Willie Johnston, John Cook, William Horsefall, and Orion Howe are a few of the boys who were awarded the Congressional Medal of Honor, and the valor of other boys was recognized in the reports of their officers.

William Bircher was a fifteen-year-old drummer boy with the 2nd Regiment Minnesota Veteran Volunteers during the war. Bircher published his diary in 1889, explaining in his preface that he never intended to write a connected story, but to make available the contents of his diary, kept during the war, to his old comrades and their families. In appearance, the contents look like a narrative, but on close examination, while it opens with a narrative about the start of the war and the beginning of the regiment's history, it becomes a chronological stream of dated entries put down in paragraph format.[38]

Charles Bardeen was a fifteen-year-old drummer boy with the 1st Massachusetts. His ancestors were keepers of diaries and he started early to do the same. His mother kept a diary for him until he was eight years old, then he took over from that time and continued throughout his life. Like many, Charles kept his diary for himself; but years after the war, a friend, Columbia University president Nicholas Murray Butler, suggested that a soldier's genuine experiences would have value. The diary was published as a book in 1910 and included over two hundred and fifty illustrations, an introduction by Butler, and a preface by Bardeen explaining the content and the art.[39]

Johnnie Wickersham is representative of those who told their story throughout life, but as the end drew near, wrote their memoirs for the family in the hope of not being forgotten. Fifty years had passed and Johnnie's recollections, according to editor, Kathleen Gorman, were flawed. His memoir was written in narrative style by chapter and topic, referenced in the back by the editor with notes and a bibliography, and indexed. The work was written in 1915 and published in 1918. Johnnie died in 1916.[40]

The conditions in the prison camps of the Civil War have been researched and published in a number of books. They have also been described by sixteen-year-old, Irish immigrant, Union soldier, Michael Dougherty. In the fall of 1863 he was captured a second time and sent to Andersonville Prison, perhaps the most notorious prison camp of the war. He kept a diary of the months he spent there through to the end of the war when the prisoners were freed and thousands were sent north by riverboat. He was with those on the steamboat *Sultana* when it exploded, and that event was also recorded in his diary. Dougherty first published his diary in 1908 in diary format with dates and entries, exactly as it was written.[41]

How and where were the stories of the experiences of these boys in the Civil War reported? Over the years, interest in their stories has waxed and waned but it began during the war when many of the boys' stories were first shared with the public through newspaper articles. Johnny Clem became an instant hero when he shot a Confederate officer off his horse as the officer tried to capture him.[42] While this article tells the story of a boy soldier, it also is an example of how a story spreads. The article is attributed to a "Cincinnati paper" and also appears in exact copy in the *Albany Evening Journal* of December 19, 1863, but with a subtitle, "He receives from Gen. Rosecrans the Badge of the Roll of Honor," also attributed to a "Cincinnati paper."[43] Not all is without controversy or error. Sometimes the error is for a purpose. Sometimes it's purely accidental, but lives on as assumed fact. When a reporter rushes to get the story to press without checking out the facts, mistakes happen. There were mistakes in the Clem article. It reported that he shot the colonel dead out of the saddle. While there were questions about this at the time, it was later learned that the colonel was wounded and recovered from his wound.[44] Newspapers were anxious to get such heroic stories to cheer up a public having a hard time dealing with the losses faced by Lincoln's armies. That played a major part in the story about Robert Hendershot whose story was told about his entrance into the battle at Fredericksburg by clinging onto the side of a boat and swimming alongside, where he captured a Confederate prisoner and lost his drum when it was blown to pieces by an artillery shell.[45]

George W. Bungay was a newspaper reporter for Horace Greeley's The Tribune Association who wrote this article which appeared in the *New Hampshire Sentinel,* as it was told to him by the boy himself, fourteen-year-old Robert Hendershot. Of particular interest is the timing as well as the point of view. The initial incident referred to in the article had taken place eleven months earlier and had already been reported by previous newspapers. This article included a history of the boy's service and the fact that the Tribune was presenting him with a brand-new drum. The boy had a habit of self- promotion and had been visiting many newspapers in the time since the original incident had made him a celebrity. What actually happened, explained Reverend George Taylor who had taken him into his care. is that he crossed the river by clinging to the stern of a boat, brought in a Confederate soldier who deserted and surrendered to Hendershot, and dropped a clock he had taken when a shell blew up nearby and surprised him.[46] Taylor went on to explained that Hendershot had given him this account of where he had been within the hearing of a number of persons including members of the press who in turn created an epic of heroism to divert the public's despondence from the devastation of the war.[47] A controversy arose after the war as some from his regiment, jealous of his fame and knowledgeable of the true facts, tried to discredit him.

Thirteen-year-old Charlie King's sad story was told in two articles; the first when he left for the war, a young patriot,[48] and the second when he returned, a victim, killed in action.[49] Another article told of the accidental death of twelve-year-old Clarence McKenzie, shot by his best friend, practicing with a gun he didn't know was loaded.[50]

Twelve-year-old Orion Howe was written up in the official records of the Medal of Honor when "…severely wounded and exposed to a heavy fire from the enemy, he persistently remained upon the field of battle until he had reported to General W. T. Sherman…"[51] A letter of recommendation from General Sherman describing that incident gained him entrance into the United States Naval Academy.[52] Willie Johnston age eleven, John Cook age fifteen, and William Horsefall age thirteen were all awarded the Medal of Honor.

The history of the boys in the Civil War was reported from the very beginning. But in the post war years, the major historical focus became the war, the generals, the places, the events, the politics. The written and photographic record about the boys gradually went into storage, becoming buried and forgotten over time.

Many collections of letters, journals, and memoirs were published after the war, some shortly after and some many years later. Most were intended for family and friends, but in many cases found their way to the general public. As veterans' groups met over the years, many wanted to put together a written record of their history and of those who served. Regimental histories were written to tell how the units were formed and by whom and to tell the history of their service. One particularly interesting history tells of a boy company in the Confederate artillery of Lee's army. *Where Men Only Dare to Go* was written by Royal Figg, an original member, to be a brief history of the Parker Battery for its veterans. But Figg decided to write a book for the general public and in so doing introduced the public to the story of the boys of the Parker Battery, many of whom were so young, as young as twelve, as to require written permission from their parents to join.[53] Published in 1885, the history of the Parker Battery became one of the first histories of boy soldiers to be released.

As the 19th century drew to a close, the histories of the Civil War began to shift. More of the historians were removed from the war by a generation. Histories became secondary sources and those writing them had to depend on the record left by those who were there. They did, however, have one major advantage, they still had access to the living veterans. They could talk to the people who had been there and lived the events of the war. Civil War histories began to appear as compilations rather than histories of individuals. In *A Brief History of the United States* by Joel and Esther Steele, the Civil War was the fifth epoch and was covered chronologically in short sections listing events, dates, and generals.[54] *History of the United States of America; for the Use of Schools*, by Charles Goodrich, was divided into six periods. Period six was distinguished for the Great Rebellion. It was an abbreviated yet wide ranging accounting of who, what, where, and when with references to cause and effect.[55] *The Centennial History of the United States* by James D. McCabe was divided into forty-five chapters, each of the first thirty representing a historic period, beginning with primitive inhabitants, with the thirty-first being the adoption of the Constitution and Washington's administration. Each of the remaining fifteen was identified by presidential administrations, ending with President Grant. Chapters forty-one and forty-two covered the Civil War by way of the two terms of President Lincoln. It was much more detailed and covered many events and incidents not mentioned in either of the two previous histories. Whereas the previous histories used headings or numerical divisions, McCabe wrote in a narrative style.[56] Cause and effect were noted and the text was enriched with photographic drawings of a woodcut style.

Nowhere in either of these histories was there any mention of the boys in the war. For example, only three sentences were allotted to the Battle of New Market in Virginia on May 15, 1864, where the Virginia Military Institute cadets fought, with the barest of information in McCabe's history.

At the beginning of the twentieth century, there was a lull in historical interest in the Civil War. Fewer regimental histories were published. The Robert Hendershot controversy came to a close, and the last veteran of the Civil War retired from active duty. Robert Hendershot was confirmed to be the drummer boy of the Rappahannock with full honors restored.[57] Susan R. Hull published her collection, *Boy Soldiers of the Confederacy,* the first collection of biographical information about boys from the war. And Johnny Clem retired as a Brigadier General after serving thirty-four years in the U.S. Army.

Susan Hull's attention was first drawn to the boys when, in 1863, Major General John Ellis Wool, commanding in Baltimore, spoke to her about revoking an order to draft all boys sixteen or older.[58] From that point on, she began collecting cuttings from newspapers with the intent to share their stories just as she received them. Her primary focus would be the stories of boys eighteen and younger. During the years following the war, Hull gathered her stories, many through correspondence to gather first-hand accounts. Her work was published in 1905. There is reference in her work of a similar effort that had already been published on behalf of Union boys, but it has not yet been located.

One of the best-known Union boys from the Civil War was John Joseph Klem who changed his name at the beginning of the war to John Lincoln Clem in honor of the president. He was nine years old when he ran away from home to join the army, twelve when he shot an enemy officer and became an instant celebrity and sergeant on Major General George Henry Thomas's staff and sixty-four when he retired a Brigadier General. He died in 1937 at age eighty-five.[59]

In the middle of the twentieth century the focus began to shift again and interest in the stories of the common soldier, and boy soldiers began to grow. In 1943 Bell Irvin Wiley's book *The Life of Johnny Reb* was published.[60] Most books, other than regimental histories and narratives, published up to that time rarely mentioned the common soldier. The men and boys who comprised the rank and file were known only through their diaries, journals, and letters where they could be found, and few were looking. But

Wiley did. Up to that point it was the biographies and histories of the generals and the events of the war that were known to most people. The common soldier came to be known -- life in camp, interests, experiences, values, heartache, and the trauma of the battlefield -- all came out from letters, diaries, and journals. Wiley read them. Then he used the words of the writers to tell their story with their own words.[61] He followed up with *The Life of Billy Yank* in 1952.

About mid-century the Civil War began to reemerge into the public awareness as an increasing number of historians began to publish their work and historic documents became available and the centenary approached. Robert Underwood Johnson and Clarence Clough Buel worked together on the *Century Magazine* beginning in 1883 to edit the reports of the officers of the Union and Confederate Armies, which originally ran in a magazine series for three years. The accounts and official reports of the battles and actions of the war, some written at the time of the war and many written after the war specifically for the series, were gathered together in a 4-volumn set, *Battles and Leaders of the Civil War*, published first by the Century Magazine in 1887 and 1888. In 1956 Thomas Yoseloff republished *Battles and Leaders* and filled it with detailed reports, maps, photographs, drawings, sketches and other art, as well as statistics covering numbers, and casualties.[62] Bruce Catton came on the scene in 1960 as senior editor for American Heritage magazine of history with *The American Heritage Picture History of the Civil War* and other works from American Heritage.[63] More Civil War diaries were published beginning in the 1950s and continuing to this day. Following are some examples.

Elisha Stockwell entered the war as a soldier at the age of fifteen. He lied a little saying he didn't know his age exactly but thought he was eighteen. Throughout the war he sent some letters home to his mother, but never kept a diary or journal. After the death of his wife in 1927 he was persuaded to try and tell the story of his war experiences for his family. At the age of eighty-one, suffering from cataracts so that he couldn't see the lines on the paper, using a piece of wood to guide his hand, Stockwell wrote his memoirs from memory. There were no chapters, few paragraphs, and little punctuation, yet the manuscript was surprisingly legible. Stockwell's daughter passed the manuscript to historian Byron Abernethy to put into readable form. Using the original manuscript, Abernathy added sentencing, paragraphing, and chapter organization to create an easy to read narrative and published Stockwell's memoirs in 1958.[64]

Val C. Giles served four years with Hood's Brigade, 4th Texas Infantry and his memoirs were published in 1961.[65] Fifteen-year-old Alfred Bellard entered the Civil War with the 5th New Jersey Infantry. His memoirs were illustrated by his own art. It was found in an attic in Pennsylvania in 1963 and published in 1975.[66] Rice C. Bull was a sergeant in the 123rd New York Volunteer Infantry. His diary described training, daily routine, and combat in the life of a soldier, and was published in 1977.[67] James M. Williams had moved south from Ohio three years before the war. When the war broke out, he joined the 21st Alabama Volunteers. Editor John Kent learned of a collection of his letters from a young university student in one of his classes and produced the surprisingly observant collection, written from a unique point of view, that of a "Northern Rebel," and published in 1981.[68] As the 20th century neared its end, the lives of the common soldier were becoming available for any who cared to look.

In 1949 *Life* magazine ran an article which pictured sixty-eight living veterans of the Civil War. Sixteen-year-old Jay S. Hoar of Rangeley, Maine, determined to meet the one who lived closest to him. On June 22, 1949, Jay took his first train and bus trip and traveled to Groff Falls, New Hampshire, where he met and visited with James M. Lurvey, a one hundred and one-year-old veteran who entered the war with his father as a fourteen-year-old drummer boy.

Lurvey recalled Gettysburg: "I never fired a shot. I was still a drummer boy[;] during much of that battle I served in the Medical Corps. Shot and shell and the screams of dying men and boys filled the humid air. A non-com told me to put away my drum. He tied a red rag around my left arm and told me I was now in the Medical Corp. I told him I was not big enough to lift my end of a stretcher, so he assigned me to a field tent. It was stifling inside. I thought I'd keel over when they told me my assignment. Wish then I could have hefted a stretcher. I was to stand by and carry out the soldiers' arms and legs as the doctor amputated them. I guess that was the day I grew up and left boyhood forever. And I wasn't yet sixteen."[69]

For Hoar, that was the day he began his life's work. In the years ahead, he juggled this work with a teaching career as he began an effort that would last over forty years to make sure that these veterans and the boys they were would not be forgotten. Correspondence, visits, travel, research led to a series of the most definitive books about the oldest veterans and the youngest to serve. *New England's Last Civil War Veterans* was published in 1976. *Callow Brave and True: A Gospel of Civil War Youth* published in 1999 was a biographical collection of the youngest, with some in the home guard as young as six and a half. *Our Eldest and Last Civil War Nurses* in 2001 was followed by *Our Youngest Blue and Gray* in 2005. A trilogy entitled *Sunset and Dusk of the Blue and the Gray* followed with volume one in 2006, two in 2008, and three in 2010.[70]

In 1989, a history honors student at the University of North Carolina, Brian Alligood, chose to investigate the youth in the war. While the vast majority of historians continued to focus on battles, campaigns, and general officers in their research and writing, and more began to be written about the common soldier participants of the war, this was a beginning of a renewed interest in the youngest soldiers.[71]

One particular event of the war was notable for the significant number of student cadets who were involved. At the Battle of New Market 250 cadets from the Virginia Military Institute fought as a unit. An exceptional collection of letters, artifacts, and biographical accounts has been gathered in the Archives of the VMI and its Hall of Valor Museum and preserved battlefield park.[72] Ten cadets lost their lives in that battle. The first to fall was seventeen-year-old William Hugh McDowell. The facts of his life and of the events of that day were skillfully woven into a remarkable book, *Ghost Cadet* by Elaine Marie Alphin.[73]

Over the next few years transitioning from the 20th into the 21st centuries two kinds of writings about boys in the Civil War were published. Anthologies with photographs, citations, quotations from original diaries and journals were published, bringing to their readers a researched collection of information about the real boys from the war. Other historians researched their subjects, then turned their stories into narrative novel format without citations, designed for younger readers to learn about the war through the eyes of their peers. As the 21st century began, more of these works became available.

Reluctant Witnesses by Emmy Werner recorded the war from Sumter to Appomattox through the words of the children who lived it – civilian and soldier, boy and girl, free and slave – gathered from diaries and journals and letters, and published in 1998. *Too Young to Die, Boy Soldiers of the Union Army 1861-1865* and *When Johnny Went Marching*, published in

2001, contained historic background and the photographs and stories of hundreds of boys from the war. *Beyond Their Years, Stories of Sixteen Civil War Children* in 2003 contained biographical information about boys and girls, black and white, civilian and soldier.[74]

Jim Murphy's *The Boys' War*, published in 1990, tells the story of the war in a narrative style comprised of the words of the boys who were there, taken from their journals, diaries, and letters.[75] It is not a record of battles and chronologies, but of the first-hand experiences of the participants. His approach to the history is unique and personal, a style of historical writing built on primary sources. After an introduction "The War Begins," the book continues to present the war by topics such as "So I Became a Soldier" and "A Long and Hungry War.". While gathering the research for *When Johnny Went Marching*, Wisler used his researched materials to write the stories of Medal of Honor winners Willie Johnston and Orion Howe, and wrote the most accurate historic fiction possible to bring to life the biographic accounts of these two boys. Willie's story, *Mr. Lincoln's Drummer* was published in 1995 and Orion's, *The Drummer Boy of Vicksburg*, in 1997.[76]

In the introduction to his book, *The Little Bugler, the True Story of a Twelve-year-old Boy in the Civil War*, published in 1998, William Styple explains his painstaking research and effort to make Gus's story as historically accurate as possible.[77] At first, he planned to write a researched account with footnotes and documentation, but he decided instead to write a narrative account in novel format so that today's youth could follow the life of Gustav Schurmann, regimental bugler to four generals and friend of Tad Lincoln. His research bibliography, acknowledgements, and picture credits are in the back of the book.

Romaine Stauffer is not a historian. She was asked by a friend to write Aaron Stauffer's story and was helped by many along the way. Aaron Stauffer was an ancestor, a boy of sixteen who went against his family and his church's Mennonite teachings and ran away to the war.[78] His story has been carefully researched through the family and the work of historian, Gary Good, in whose book *Faith, Hope, and Love* can be found Aaron's Civil War records and those of others with whom he ran away. *Aaron's Civil War* was published in 2011 in the form of a novel with a research bibliography in the back.[79]

Over one hundred and fifty years have passed and the tens of thousands of boys and youth who once trod the field of battle are long gone. Their record remains. Like all of history, it's there for all to see if we choose to look. They were their own historians as they shared their wartime experiences, their thoughts and hopes and dreams and fears in their journals, their diaries, their letters. Their deeds were recorded in the newspapers of the day, in citations in dusty archives, in the records of their lives written by those who knew them, or friends, or family. As the years passed on, the writing of their story changed. Others recorded what they remembered. In later years, the boys grown up forgot the details or embellished their part in what happened. Historians of later generations have tried to gather the events and the details to record that which in some cases had only been told to family but never written down. In some cases, there is a treasure trove of information, carefully preserved and recorded. Probably the largest collection of documents about the boys of the Civil War is the archive collection at the Virginia Military Institute. In 1991, Susan Provost Beller published her book, *Cadets at War, The True Story of Teenage Heroism at the Battle of New Market*, telling the story of what happened that day and the days that followed, and of many of the individual cadets who took part in the battle.[80] It was based on the primary sources in the library at VMI including maps, photographs, and written accounts and reports. In other cases, historians such as Susan R. Hull, Jay Hoar, Scotti Cohn, Jim Murphy, Clifton Wisler, and Dennis Keesee have made it their mission to gather the stories into collections for others to read and remember that Johnny, Robert, William, Orion, and Willie and all the rest, did pass this way and are a part of our history.

This is a history of the history of the boys of the Civil War. It began with them and has evolved over the years. Along the way, many have helped to share their stories and keep their memory alive. In recent years, several historians have continued to work to research, preserve, and share their history, including the last historian to have had personal contact with the last surviving boy soldiers.

ENDNOTES

1   Emmy E. Werner, Reluctant Witnesses, Children's Voices from the Civil War (Boulder CO: Westview Press, 1998), 2.

2   Jay S. Hoar, Callow, Brave, and True, a Gospel of Civil War Youth (Gettysburg, PA: Thomas Publications, 1999), xvii, xix. At the start of the Civil War, the presence of young boys in the ranks was taken pretty much for granted. Their presence extended historically to Old World tradition with many examples including fourteen-year-old Donald McCloud in 1702, eleven-year-old James Christian in 1703, twelve-year-old Marquis de Montcalm in 1724, and many others over the years. This trend continued throughout early American warfare. Examples included thirteen-year-old Eleazar Clay in the French and Indian War, fourteen-year-old Elijah Kellogg, Sr. in the American Revolution, eight-year-old Roswell Woolson in the War of 1812, and seven-year-old James Buckner in the Mexican War.

3   Dennis M. Keesee, Too Young to Die, Boy Soldiers of the Union Army 1861-1865 (Huntington, WV: Blue Acorn Press, 2001), 7-11.

4   Jim Murphy, The Boys' War (New York: Clarion Books, 1990), 10.

5   Keesee, Too Young, 25.

6   Hoar, Callow, 222-3.

7   G. Clifton Wisler, When Johnny Went Marching (New York: Harper-Collins, 2001), viii.

8   Bell Irvin Wiley, The Life of Johnny Reb (Baton Rouge: Louisiana State University Press, 1943), 333. Foremost among these schools were the Virginia Military Institute in Lexington, Virginia, Hillsboro Military Academy in Hillsboro, North Carolina, and the Citadel in Charleston, South Carolina. Perhaps the youngest instructor to enter service in the Confederate army this way was eleven-year-old Charles Carter Hay. (Hoar, Callow, 52-55).

9   Wisler. When Johnny Went Marching, viii.

10   Elisha Stockwell, Private Elisha Stockwell, Jr. Sees the Civil War, Bryon R. Abenethy, ed. (Norman: University of Oklahoma Press, 1985), 5-6.

11   Richard Bak, "Michigan's Little Drummer Boys of the Civil War," Hour Detroit, December 2011, 5, http://www.hourdetroit.com/Hour-Detroit/December-2011/Rhythm-Section-Civil-War-Sesquicentennial/ , accessed December 8, 2016.

12   Hoar, Callow, 177-8.

13   Gary Good, Glaube, Hoffnung, und Liebe; Faith, Hope, and Love (Morgantown, PA: Masthof Press, 1996), 72.

14   William B. Styple, The Little Bugler, The True Story of a Twelve-Year-Old Boy in the Civil War (Kearny, NJ: Belle Grove Publishing, 1998), 16-18.

15   Hoar, Callow, 90-91.

16   Emmy E. Werner, Reluctant Witnesses, Children's Voices from the Civil War, (Boulder CO: Westview Press, 1998), 9.

17   Ibid., 8.

18   William Bircher, A Drummer Boy's Diary: Four Years of Service with the Second Regiment Minnesota Veteran Volunteers 1861 to 1865, (St. Paul, MN: St. Paul Book and Stationery Company, 1889), 9-36.

19   Johnnie Wickersham, Boy Soldier of the Confederacy, The Memoir of Johnnie Wickersham. Kathleen Gorman, ed. (Carbondale: Southern Illinois University Press, 2006), 16-17.

20   Murphy, Boys' War, 29.

21   Keesee, Too Young, 11-12.

22   Ibid., 29,30.

23   Ibid., 149.

24   Ibid., 148-149.

25   Photos and information donated by Jody Clevenger a descendent of Brice E. Davis. https://www.flickr.com/photos/civilwar_veterans_tombstones/3208238232/ , accessed December 8, 2016.

26   Michael Dougherty, Prison Diary of Michael Dougherty (Bristol, PA: Charles A. Dougherty, printer, 1908).

27   Keesee, Too Young, 152-4.

28   Wiley, Johnny Reb, 322-34.

29   Brian Alligood, "Boys in Gray: The Role of Confederate Youth in the American Civil War," Honors Essay, Department of History, University of North Carolina (1989): 13-15.

30   Bell Irvin Wiley, The Life of Billy Yank. (Baton Rouge: Louisiana State University Press, 1952), 296- 302.

31   Werner, Reluctant Witnesses, 2.

32   Ibid., 3.

33   James Marten, ed., Children and Youth during the Civil War, (New York: New York University Press, 2012), 4.

34   Werner, Reluctant Witnesses, 3.

35   Ibid., 81.

36   A statue depicting Sergeant Kirkland, known the Angel of Mayre's Heights, is in front of the Stone Wall at Mayre's Heights. Kirkland was later promoted to Second Lieutenant and was killed in September 1863 at the battle of Chickamauga.

37   Ibid., 156, 84.

38   William Bircher, A Drummer Boy's Diary: Four Years of Service with the Second Regiment Minnesota Veteran Volunteers 1861 to 1865, (St. Paul, Minnesota: St. Paul Book and Stationery Company, 1889), 5-6.

39   Charles W. Bardeen, A Little Fifer's War Diary (Syracuse, New York: C. W. Bardeen, 1910), 5-9.

40   Johnnie Wickersham, Boy Soldier of the Confederacy, The Memoir of Johnnie Wickersham, Kathleen Gorman, ed., (Carbondale: Southern Illinois University Press, 1918).

41   Michael Dougherty, Prison Diary of Michael Dougherty (Bristol, Pennsylvania.: Charles A. Dougherty, printer, 1908), from page 122 April 23rd,1865 – Vicksburg, Miss. Went aboard the boat called the "Sultana" to be taken to St. Louis, Mo. There are about 2,200 of us, mostly old prisoners from Andersonville, Ga. On the trip up the Mississippi, the "Sultana" met with a terrible disaster [its boilers exploded], causing complete destruction of the boat; and hundreds of men who had passed safely through many bloody battles and the horrible suffering of Southern prison life perished within but a few days' journey of home and friends.

42   Michael Dougherty, "The Youngest Soldier in the Army of the Cumberland," National Aegis, December 26, 1863, The Family Circle, 1.

43   Michael Dougherty, The Youngest Soldier in the Army of the Cumberland," Albany Evening Journal, December 19, 1863, 2.

44   Dennis M. Keesee, Too Young to Die, Boy Soldiers of the Union Army 1861-1865 (Huntington, West Virginia: Blue Acorn Press, 2001), 231.

45   George W. Bungay, "The Drummer Boy of the Rappahannock." New Hampshire Sentinel, November 19, 1863, 1.

46   Richard Bak, "Michigan's Little Drummer Boys of the Civil War," Hour Detroit, December 2011, http://www.hourdetroit.com/Hour-Detroit/December-2011/Rhythm-Section-Civil-War-Sesquicentennial/ accessed December 8, 2016.

47   Anthony Patrick Glesner, America's Civil War 16 No. 6 (January 2004): 26.

48   "Young Patriotism," Village Record (West Chester, Pennsylvania), December 31, 1861.

49   "Obituary of Charles King," Village Record (West Chester, Pennsylvania), October 2, 1862.

50   "The Death of Young McKenzie," Brooklyn Eagle, War Intelligence, June 13, 1861.

51   United States Army Center of Military History Website/Medal of Honor/Civil War, http://www.history.army.mil/moh/civilwar_gl.html#top , accessed December 8, 2016.

52   Frank Moore, The Civil War in Song and Story 1860-1865 (New York: P. F. Collier, Publisher, 1889), 104

53   Royall W. Figg, "Where Men Only Dare to Go!" Or the Story of a Boy Company C.S.A., By an Ex-Boy (Richmond, Virginia: Whittet & Shepperson, 1885), vii-viii.

54   Joel Dorman Steele and Esther Baker Steele, A Brief History of the United States (New York: American Book Company, 1885), 214-80.

55   Charles A. Goodrich, History of the United States of America; for the Use of Schools (Boston: Brewer and Tillston, 1876), 238-309.

56   James D. McCabe, The Centennial History of the United States, From the Discovery of the American Continent to the Close of the First Century of American Independence (Philadelphia: The National Publishing Company, 1874), 779-864.

57   Anthony Patrick Glesner, "The Drummer Boy of the Rappahannock," America's Civil War, January 2004. http://www.historynet.com/americas-civil-war-drummer-boy-of-the-rappahannock.htm, accessed December 8, 2016.

58   Susan R. Hull, collated by, Boy Soldiers of the Confederacy, 1998 Eakin Press ed. (New York: Neale Publishing Company, 1905), 13-14.

59   "John Clem." Wikipedia, the Free Encyclopedia. http://en.wikipedia.org/wiki/John_Clem, accessed December 8, 2016.

60   Bell Irvin Wiley, The Life of Johnny Reb (Baton Rouge: Louisiana State University Press, 1943).

61   C. E. Dornbusch, comp., Military Bibliography of the Civil War, 4 vols. (New York: The New York Public Library, 1972), 3:127.

62   Robert Underwood Johnson and Clarence Clough Buel, eds., "The Century Magazine," Battles and Leaders of the Civil War, 4 vols. (New York: Century Company, 1887-1888).

63   Richard M. Ketchum, ed. in charge and Bruce Catton, narrative, The American Heritage Picture History of the Civil War, (New York: American Heritage Publishing Company, 1960).

64   Elisha Stockwell, Private Elisha Stockwell, Jr. Sees the Civil War, Bryon R. Abernethy, ed. (Norman: University of Oklahoma Press, 1985), ix-xii.

65   Val C. Giles, Rags and Hope, The Recollections of Val C. Giles (New York: Coward-McCann, 1961).

66   Alfred Bellard, Gone for a Soldier: The Civil War Memoirs of Private Alfred Bellard, David Herbert Donald, ed. (Boston: Little, Brown, 1975).

67   Rice C. Bull, Soldiering - The Civil War Diary of Rice C. Bull, K. Jack Bauer, ed. (San Rafael, CA: Presidia Press, 1977).

68   James M. Williams, From that Terrible Field, Civil War Letters of James M. Williams, John Kent Folmar, ed. (Tuscaloosa: University of Alabama Press, 1981).

69   Clayton, John, at large, "New Hampshire's Last Boy in Blue Lives on in Legend," New Hampshire Union Leader (Manchester, New Hampshire: Joseph McQuaid, July 3, 1998),16.

70   Jay S. Hoar, New England's Last Civil War Veterans (Arlington, VA: Seacliff Press, 1976): Jay S. Hoar, Callow, Brave, and True: A Gospel of Civil War Youth (Gettysburg, PA: Thomas Publications, 1999); Jay S. Hoar, Our Eldest and Last Civil War Nurses (Temple, MN: Bo-ink-um Press, 2001); Jay S. Hoar, Our Youngest Blue and Gray, Callow Brave and True (Salem, MA: Higginson Book Company, 2005); Jay S. Hoar Sunset and Dusk of the Blue and the Gray: Last Living Chapter of the American Civil War. An Epic Prose Elegy, 3 vols. (Salem, MA: Higginson Book Company, 2006, 2007, 2010).

71   Brian Alligood, "Boys in Gray: The Role of Confederate Youth in the American Civil War," Honors essay: Department of History, University of North Carolina, 1989.

72   Archives of the Virginia Military Academy, http://www.vmi.edu/archives/home/, accessed December 8, 2016.

73   Elaine Marie Alphin, Ghost Cadet (Princeton, Illinois: Hither Page Press, 1991).

74   Emmy E. Werner, Reluctant Witnesses, Children's Voices from the Civil War (Boulder CO: Westview Press, 1998); Dennis M. Keesee, Too Young to Die, Boy Soldiers of the Union Army 1861-1865 (Huntington, WV: Blue Acorn Press, 2001); G. Clifton Wisler, When Johnny Went Marching (New York: Harper Collins, 2001); Scott Cohn, Beyond Their Years, Stories of Sixteen Civil War Children (Guilford, CT: The Globe Pequot Press, 2003).

75   Jim Murphy, The Boys' War (New York: Clarion Books, 1990).

76   G. Clifton Wisler, Mr. Lincoln's Drummer (New York: Dutton Children's Books, 1995); G. Clifton Wisler, The Drummer Boy of Vicksburg (New York: Lodestar Books, 1997).

77   William B. Styple, The Little Bugler, The True Story of a Twelve-Year-Old Boy in the Civil War, (Kearny, New Jersey: Belle Grove Publishing Company, 1998).

78   Romaine Stauffer, Aaron's Civil War (Harrisonburg, VA: Christian Light Publications, 2011);

79   Gary Good, Glaube, Hoffnung, und Liebe; Faith, Hope, and Love (Morgantown, Pennsylvania: Masthof Press, 1996).

80   Susan Provost Beller, Cadets at War: The True Story of Teenage Heroism at the Battle of New Market, (Lincoln, NE: iUniverse.com, 2000).

BOYS OF THE CIVIL WAR
SOURCE MATERIAL
By J. Arthur Moore

If you can read only one book

Author	Title. City: Publisher, Year.
Keesee, Dennis M.	*Too Young to Die, Boy Soldiers of the Union Army 1861–1865*. Huntington, WV: Blue Acorn Press, 2001.

Books and Articles Primary Sources

Author	Title. City: Publisher, Year. | "Title," in Journal ##, no. # (Date): #.
Bardeen, Charles W.	*A Little Fifer's War Diary*. Syracuse, NY: C. W. Bardeen, 1910.
Bauer, K. Jack, ed.	*Soldiering—The Civil War Diary of Rice C. Bull*. San Rafael, CA: Presidia Press, 1977.
Bellard, Alfred	*Gone for a Soldier: The Civil War Memoirs of Private Alfred Bellard*. David Herbert Donald, ed. Boston: Little, Brown, 1975.
Bircher, William	*A Civil War Drummer Boy, Diary of William Bircher, 1861–1865*. Shelley Swanson, ed. Mankato, MN: Blue Earth Books, 1999.
Bungay, George W.	"The Drummer Boy of the Rappahannock." *New Hampshire Sentinel*, November 19, 1863.
Dougherty, Michael	*Prison Diary of Michael Dougherty*. Bristol, PA: Charles A. Dougherty, 1908.

Figg, Royall W.	*"Where Men Only Dare to Go!" Or the Story of a Boy Company C.S.A.*, By an Ex-Boy. Richmond, Virginia: Whittet & Shepperson, 1885.
Giles, Val C.	*Rags and Hope, The Recollections of Val C. Giles.* New York: Coward-McCann, 1961.
Hull, Susan R., collated by	*Boy Soldiers of the Confederacy*, New York: Neale Publishing, 1905.
Moore, Frank	*The Civil War in Song and Story 1860-1865.* New York: P. F. Collier, 1889.
Schiller, Herbert M., ed.	*A Captain's War, the Letters and Diaries of William H. S. Burgwyn 1861-1865.* Shippensburg, PA: White Mane Publishing, 1994.
Wickersham, Johnnie	*Boy Soldier of the Confederacy, The Memoir of Johnnie Wickersham.* Kathleen Gorman, ed. Carbondale: Southern Illinois University Press, 1918.
Williams, James M.	*From that Terrible Field, Civil War Letters of James M. Williams.* John Kent Folmar, ed. Tuscaloosa: University of Alabama Press, 1981.
	"The Death of Young McKenzie." *Brooklyn Eagle*, War Intelligence, June 13, 1861.
	"Young Patriotism." *Village Record*, (West Chester, PA) December 31, 1861.
	Obituary Charles King. *Village Record* (West Chester, PA), October 2, 1862.
	"The Youngest Soldier in the Army of the Cumberland." *National Aegis*, December 26, 1863, The Family Circle.

Secondary Sources

Author	Title. City: Publisher, Year. | "Title," in Journal ##, no. # (Date): #.
Alligood, Brian	"Boys in gray: the role of Confederate youth in the American Civil War." Honors essay, Department of History, University of North Carolina, 1989.
Alphin, Elaine Marie	Ghost Cadet. Princeton, IL: Hither Page Press, 1991.
Bower, Bert and Jim Lobdell	History Alive! America's Past. Palo Alto, California: Teacher's Curriculum Institute, 2003.
Cohn, Scotti	Beyond Their Years, Stories of Sixteen Civil War Children. Guilford, Connecticut: The Globe Pequot Press, 2003.
Clayton, John	Joseph McQuaid, pub. "New Hampshire's Last Boy in Blue Lives on in Legend." New Hampshire Union Leader, July 3, 1998.
Davidson, James West, Pedro Castillo, and Michael B. Stoff	The American Nation. Upper Saddle River, NJ: Prentice Hall, 2002.
Davis, Archie K.	Boy Colonel of the Confederacy: The Life and Times of Henry King Burgwyn, Jr. Chapel Hill: University of North Carolina Press, 1985.
Davis, James A.	"Union Musicians and the Medal of Honor During the American Civil War," College Music Symposium 54 (July 8, 2014).
Dornbusch, C. E., comp.	Military Bibliography of the Civil War, vol. 3. New York: The New York Public Library, 1972.
Drago, Edmund	Confederate Phoenix: Rebel Children and Their Families in South Carolina. New York: Fordham University Press, 2008.
Fox, Thomas	Drummer Boy Willie McGee, Civil War Hero and Fraud. Jefferson, NC: McFarland, 2008.

Good, Gary	*Glaube, Hoffnung, und Liebe; Faith, Hope, and Love*. Morgantown, PA: Masthof Press, 1996.
Hoar, Jay S.	*New England's Last Civil War Veterans*. Arlington, TX: Seacliff Press, 1976.
———.	*Callow, Brave, and True, a Gospel of Civil War Youth*. Gettysburg, PA: Thomas Publications, 1999.
———.	*Our Eldest and Last Civil War Nurses*. Temple, MN: Bo-ink-um Press, 2001.
———.	*Montana's Last Civil War Veterans*. Temple, MN: Bo-ink-um Press, 2010.
———.	*Our Youngest Blue and Gray: A Gospel of Civil War Youth*. Salem, MA: Higginson Book Company, 2005.
———.	*Sunset and Dusk of the Blue and the Gray: Last Living Chapter of the American Civil War. An Epic Prose Elegy*. Vol. 1, *The North's Last Boys in Blue*. Salem, ME: Higginson Book Company, 2006.
———.	*Sunset and Dusk of the Blue and the Gray: Last Living Chapter of the American Civil War. An Epic Prose Elegy*. Vol. 2, *The North's Last Boys in Blue*. Salem, ME: Higginson Book Company, 2008.
———.	*Sunset and Dusk of the Blue and the Gray: Last Living Chapter of the American Civil War. An Epic Prose Elegy*. Vol. 3, *The South's Last Boys in Gray*. Salem, ME: Higginson Book Company, 2010.
Howe, Henry	*Historical Collections of Ohio*, vol. 2. Cincinnati: State of Ohio/Laning Printing Company, 1896.
Jabour, Anya	*Topsy-Turvy: How the Civil War Turned the World Upside Down for Southern Children*. Lanham, MD: Ivan R. Dee/Rowman & Littlefield, 2010.

Marten, James	*Children for the Union: The War Spirit on the Northern Home Front.* Chicago, IL: Ivan R. Dee, 2004.
Marten, James, ed.	*Children and Youth during the Civil War.* New York: New York University Press, 2012.
Murphy, Jim	*The Boys' War.* New York: Clarion Books, 1990.
Robertson, Ellen	"Major General John Lincoln Clem," *On Point: Journal of Army History* 19, no. 2 (Fall 2013): 18-21.
Smith, Timothy B.	"Myths of Shiloh," in *America's Civil War*, 19, no. 5 (May 2006): 30-71.
Stauffer, Romaine	*Aaron's Civil War.* Harrisonburg, Virginia: Christian Light Publications, 2011.
Steele, Joel Dorman and Esther Baker Steele	*A Brief History of the United States.* New York: American Book Company, 1885.
Styple, William B.	*The Little Bugler, The True Story of a Twelve-Year-Old Boy in the Civil War.* Kearny, NJ: Belle Grove Publishing Company, 1998.
Why We Remember United States History. New York: Addison-Wesley Publishing Company, 1998.	*Why We Remember United States History.* New York: Addison-Wesley Publishing Company, 1998.
Reluctant Witnesses, Children's Voices from the Civil War. Boulder Colorado: Westview Press, 1998.	*Reluctant Witnesses, Children's Voices from the Civil War.* Boulder Colorado: Westview Press, 1998.
The Life of Johnny Reb. Baton Rouge: Louisiana State University Press, 1943.	*The Life of Johnny Reb.* Baton Rouge: Louisiana State University Press, 1943.
The Life of Billy Yank. Baton Rouge: Louisiana State University Press, 1952.	*The Life of Billy Yank.* Baton Rouge: Louisiana State University Press, 1952.

Mr. Lincoln's Drummer. New York: Dutton Children's Books, 1995.	*Mr. Lincoln's Drummer.* New York: Dutton Children's Books, 1995.
The Drummer Boy of Vicksburg. New York: Lodestar Books, 1997.	*The Drummer Boy of Vicksburg.* New York: Lodestar Books, 1997.
When Johnny Went Marching. New York: Harper-Collins, 2001.	*When Johnny Went Marching.* New York: Harper-Collins, 2001.

Web Resources

URL	Name and description
http //histclo.com/ youth/uncw.html	Children in History is a website devoted to developing information about children in history. It has material on the Civil War including a section on The Boys' War. Although the page asks for a log in when first launched it can be accessed by simply clicking Cancel on the log in screen.
http://www.historynet. com/americas-civil-war- drummer-boy-of-the- rappahannock.htm	Glesner, Anthony Patrick. "The Drummer Boy of the Rappahannock." *America's Civil War*, January 2004.
http://www.pbs.org/wgbh/ americanexperience/features/ general-article/grant-kids/	"Kids in the Civil War." general article, WGBH American Experience PBS.
http://www.history.army. mil/moh/civilwar_af.html	The United States Army Center of Military History Website lists Civil War Medal of Honor recipients.

Other Sources

Name	Description, Contact information including address, email
Virginia Military Institute Civil War and New Market	The VMI Civil War archives contain a great deal of information about boys in the Civil War with a catalogue available on line but most of the materials only accessible on site. Their website is: http://www.vmi.edu/archives/civil-war-and-new-market/
"John Clem," Wikipedia	This is the Wikipedia entry on John Lincoln Clem. https://en.wikipedia.org/wiki/John_Clem

Scholars

Name	Email
J. Arthur Moore	Joemoore3@comcast.net
Jay S. Hoar	deceased

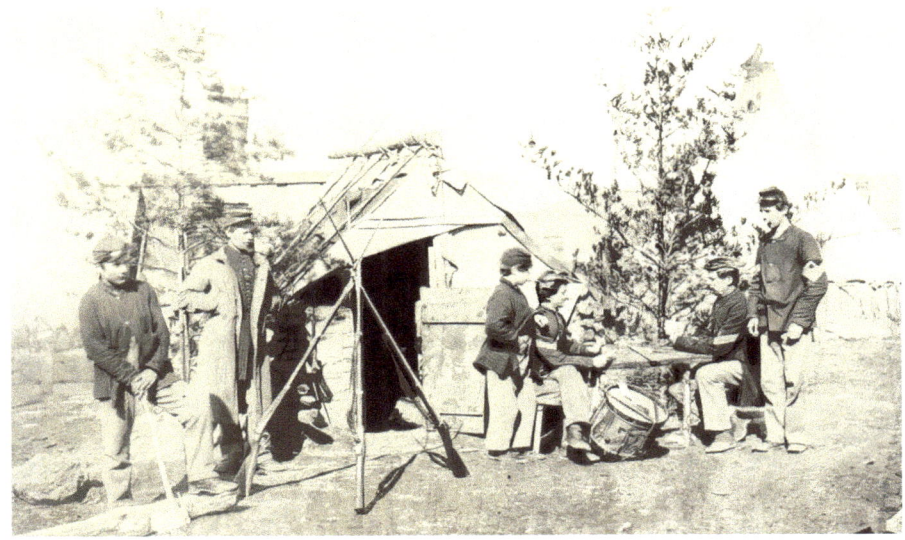

Winter 1862

The War from the Boys' Point of View

Unlike later wars in American history, young people were involved in all aspects of the Civil War, including fighting on the battlefield. William Black, one of the youngest wounded soldiers, was twelve when his left hand and arm were shattered by an exploding shell. An unknown number of soldiers—probably around five percent—were under eighteen, and some were as young as ten. Other boys and girls served as scouts or nurses for the wounded. Yet even those who did not participate in the war itself saw their lives altered by the conflict. During wartime, young people had to grow up quickly, assuming the responsibilities of absent relatives.

ENLISTING

In 1861, President Lincoln announced that boys under eighteen could enlist only with their parents' consent. The next year, he prohibited any enlistment of those under eighteen. But heavy casualties led recruiting officers to look the other way when underage boys tried to enlist, and thousands participated in the conflict as drummers, messengers, hospital orderlies, and often as full fledged soldiers. They carried canteens, bandages, and stretchers, and assisted surgeons and nurses. Many young soldiers signed up as drummers, who relayed officers' commands, signaling reveille, roll call, company drill, and taps. In the heat of battle, many carried orders or assisted with the wounded; at least a few picked up rifles and participated in the fighting.

Their motives for enlisting varied, including patriotism and a desire to escape the boring routine of farm life or an abusive family. A few were jealous of older brothers, and some young Northerners were eager to rid the country of slavery. For some young Confederates, there was a desire to repel northern invaders from their soil. One southern boy made his feelings clear with words colored by irony: "I rather die then be com a Slave to the North."

Elisha Stockwell of Alma, Wisconsin, was fifteen years old when he enlisted.

"We heard there was going to be a war meeting at our little log school house. I went to the meeting when they called for volunteers, Harrison Maxon (21), Edgar Houghton (16), and myself, put our names down.... My father was there and objected to my going, so they scratched my name out, which humiliated me somewhat. My sister gave me a severe calling down...for exposing my ignorance before the public, and called me a little snotty boy, which raised my anger. I told her, 'Never mind, I'll go and show you that am not the little boy you think I am.'

"The Captain got me in by lying a little, as I told the recruiting officer I didn't know just how old I was but thought I was eighteen. He didn't measure my height, but called me five feet five inches high. I wasn't that tall two years later when I re-enlisted, but they let it go, so the records show that as my height.

"I told her [his sister] I had to go down town. She said, "Hurry back, for dinner will soon be ready." But I didn't get back for two years."

Elisha Stockwell, quoted in Jim Murphy, The Boys' War, 13, 14

DRILLING

The Union Army was unprepared for a major war, as some young soldiers quickly discovered.

"There was considerable delay in issuing us clothing and equipment. It was not until the second week of [1861] that we were issued wooden guns, wooden swords and cornstalks with which to drill and mount guard. We went to parade in our shirts, still not being fully uniformed."

Thomas Galwey of the Eighth Ohio Regiment, quoted in Emmy E. Werner, Reluctant Witnesses, 12

MARCHING

Excitement over enlistment swiftly gave way to the boring routines of camp life and marches.

"Day after day and night after night did we tramp along the rough and dusty roads 'neath the most broiling sun with which the month of August ever afflicted a soldier; thro' rivers and their rocky valleys, over mountains—on, on, scarcely stopping to gather the green corn from the fields to serve us for rations.... During these marches the men are sometimes unrecognizable on account of the thick coverings of dust which settle upon their hair, eyebrows and beard, filling likewise the mouth, nose, eyes, and ears."

Sixteen-year-old Confederate soldier John Delhaney, quoted in Murphy, Boys' War, 27

FIGHTING

Young soldiers' romantic illusions about military glory evaporated under the harsh realities of combat. They suffered hunger, fatigue, and discomfort, and gradually lost their innocence in combat. Every aspect of soldiering comes alive in their letters and diaries: the stench of spoiled meat, the deafening sound of cannons, the sight of maimed bodies, and the randomness and anonymity of death.

"As we lay there and the shells were flying over us, my thoughts went back to my home, and I thought what a foolish boy I was to run away to get into such a mess I was in. I would have been glad to have seen my father coming after me."

Elisha Stockwell after the battle of Shiloh in Tennessee in 1862. Quoted in Murphy, Boys' War, 33

"The rains have uncovered many of the shallow graves. Bony knees, long toes, and grinning skulls are to be seen in all directions. In one place I saw a man's boot protruding from the grave…leaving the skeleton's toes pointing to a land where there is no war."

Thomas Galwey, quoted in Werner, Reluctant Witnesses, 17

"I passed . . . the corpse of a beautiful boy in gray who lay with his blond curls scattered about his face and his hand folded peacefully across his breast. He was clad in a bright and neat uniform, well garnished with gold, which seemed to tell the story of a loving mother and sisters who had sent their household pet to the field of war. His neat little hat lying beside him bore the number of a Georgia regiment …. He was about my age …. At the sight of the poor boy's corpse, I burst into a regular boo hoo and started on."

John A. Cockerill, 16, Union regimental musician, at Pittsburg Landing, Mississippi, April 1862, quoted in Emmy E. Werner, Reluctant Witnesses, 25

". . . I was certainly scared. One shell had exploded near enough so that I could realize its effects, and the one thing I wanted was to get where no more shells could burst around me. This patriotic hero who had declared in front of campfires how he had longed for gore would have liked to be tucked up once more in his little trundle bed. Bomb ague is a real disease and I had caught it.

"There was no question of getting back to the regiment …. I could see that my division was preparing to march, and while I did not actually run I certainly walked fast to get to it. It is curious how little annoyances will keep themselves prominent even in time of danger. I had on thick woolen drawers which had somehow broken from the fastening that held them up. It was a warm day and as I hurried up the hill those drawers kept slipping down till they drove me almost distracted, disturbing my equanimity more than the danger did."

Charles W. Bardeen, a fifteen year old drummer boy with the First Massachusetts Regiment, at Fredericksburg, Virginia, in December, 1862, A Little Fifer's War Diary, 107

"Dear Mother,

My first battle is over and I saw nearly all of it.... Saturday the hardest fighting was done. I saw the Irish Brigade make three charges. They started with full ranks, and I saw them, in less time than it takes to write this, exposed to a galling fire of shot and shell and almost decimated.... I saw wounded men brought in by the hundred and dead men lying stark on the field, and then I saw our army retreat to the very place they started from, a loss incalculable in men, horses, cannon, small arms, knapsacks, and all the implements of war, and I am discouraged. I came out here sanguine as any one, but I have seen enough, and I am satisfied that we never can whip the South.... Let any one go into the Hospital where I was and see the scenes that I saw...."

Charles W. Bardeen, quoted in Werner, Reluctant Witnesses, 36

"The sight of hundreds of prostrate men with serious wounds of every description was appalling. Many to relieve their suffering were impatient for their turn upon the amputation tables, around which were pyramids of severed legs and arms.... Many prayed aloud, while others shrieked in the agony and throes of death."

Edward W. Spangler, a sixteen year old with the 130th Pennsylvania Regiment, at the battle of Antietam in 1862. Quoted in Werner, Reluctant Witnesses, 32

"September 19th: Hot and dusty. At daybreak, as we marched along, we saw troops falling into line on the right of the road; the artillery was unlimbered, the gunners stood to their guns, and every thing had the appearance of a battle. We marched along the rear of the line until we reached the left wing of the army, where we piled up our knapsacks, formed in line, marched to the front, and deployed skirmishers. We advanced but a short distance in the woods, which was a pine forest, before we came upon the rebel skirmish-line. We heard on our right the heavy roll of musketry and the terrible thunder of the artillery, and it came nearer and nearer, until, in less time than it takes to describe it, we were engaged with Bragg's army. The terrible carnage continued at intervals all day.

"At night we heard, from all over the field, the cry of the wounded for water and help, and the ambulance corps were doing all in their power to bring all the wounded into our lines. The night was cool, with a heavy frost, and the water was very scarce. We lay on our arms all night, and on Sunday, the 20th, the battle was renewed with terrible slaughter on both sides. Towards noon we heard that Chittenden's and McCook's corps, on our right, had been driven back, and all that was left on the field, to hold in check the entire rebel army, was our corps,—Thomas's Fourteenth. We held the enemy back until evening, in spite of his desperate assaults, and after dark we retired to Rossville. Here General Thomas posted Negley's right, stretching to the Dry Valley Road, Brannan's (our) division in reserve to Reynolds's right and rear, while McCook's corps extended from Dry Valley nearly to Chattanooga Creek. Bragg's army was too tired and too sadly worsted to attempt to follow on the night of the 20th. On the 21st a few straggling shots were directed against our army at Rossville. Thomas felt that he could not hold his position there against the Confederate army. Orders were received at 6 p.m. on the 21st, and by seven o'clock the next morning our army was withdrawn, without opposition from the enemy. This ended the battle of Chickamauga. Though retiring from the field, our army had succeeded in shutting the rebels out of Chattanooga."

From "A Drummer Boy's Diary: Comprising Four Years of Service with the Second Regiment Minnesota Veteran Volunteers, 1861-1865, drummer boy William Bircher recounts his observations and experiences at the commencement of the Battle of Chickamauga.

LIFE AS A SOLDIER

Young soldiers frequently complained about a lack of equipment, inadequate clothing, and the quality of the food.

"After we had been in the field a year or two the call, 'Fall in for your hard-tack!' was leisurely responded to by only about a dozen men…. Hard-tack was very hard. This I attributed to its great age, for there was a common belief among the boys that our hard-tack had been baked long before the beginning of the Christian era. This opinion was based upon the fact that the letters "B.C." were stamped on many, if not, indeed, all the cracker-boxes."

Fifteen-year-old William Bircher of St. Paul, Minnesota, A Drummer-boy's Diary: Comprising Four Years of Service with the Second Regiment Minnesota

"Again we sat down beside [the campfire] for supper. It consisted of hard pilot-bread, raw pork and coffee. The coffee you probably would not recognize in New York. Boiled in an open kettle, and about the color of a brownstone front, it was nevertheless…the only warm thing we had. The pork was frozen, and the water in the canteens solid ice, so we had to hold them over the fire when we wanted a drink. No one had plates or spoons, knives or forks, cups or saucers. We cut off the frozen pork with our pocket knives, and one tin cup from which each took a drink in turn, served the coffee."

Sixteen-year-old Charles Nott of New York, quoted in Murphy, Boys' War, 48-49

"We managed to find four blankets, two of them wet and frozen, and a buffalo skin. The snow was scraped away from the windward side of the fire, and the frozen blankets were laid on the ground – a log was rolled up for a wind-break, and the buffalo [skin] spread over the blankets. On this four of us were stretched, and very close and straight we had to lie."

Charles Nott, quoted in Murphy, Boys' War, 55

"We marched through Corinth [Mississippi] in a cold, drizzly rain, and as I didn't have my blankets, I was wet through. I suffered that night as we had only green wood to make a fire. It stopped raining so I got my clothes partly dried. I lay down on the wet ground to sleep, but would get so cold that I would have to get up and hover over the smoky fire. I put in about the most disagreeable night in my life."

Elisha Stockwell, quoted in Murphy, Boys' War, 56

CONFINEMENT IN A CONFEDERATE PRISON CAMP

The accounts of young Union prisoners at Confederate prison camps are especially harrowing. Sixteen-year-old Michael Dougherty was shocked by the sight of "different instruments of torture: stocks, thumb screws, barbed iron collars, shackles, ball and chain. Our prison keepers seemed to handle them with familiarity." William Smith, a fifteen-year-old soldier in the 14th Illinois Infantry, was shaken by the physical appearance of prisoners at Andersonville in Georgia, a "great mass of gaunt, unnatural-looking beings, soot-begrimes, and clad in filthy trousers."

Michael Dougherty was the only member of his company to survive imprisonment at Andersonville Prison in Georgia.

"No one, except he was there in the prison can form anything like a correct idea of our appearance about this time. We had been in prison nearly five months and our clothing was worn out. A number were entire naked; some would have a ragged shirt and no pants; some had pants and no shirt; another would have shoes and a cap and nothing else. Their flesh was wasted away, leaving the chaffy, weather beaten skin drawn tight over the bones, the hip bones and shoulders standing out. Their faces and exposed parts of their bodies were covered with smoky black soot, from the dense smoke of pitch pine we had hovered over, and our long matted hair was stiff and black with the same substance, which water would have no effect on, and soap was not to be had. I would not attempt to describe the sick and dying, who could now be seen on every side."

Michael Dougherty, who was 16 when he joined the 13th Pennsylvania Cavalry, Diary of a Civil War Hero, p. 43.

THE YOUNGEST TO SERVE
DURING THE CIVIL WAR

six of the more than 200,000 who served

Willie Johnston, 12
Drummer, D Co. 3rd Vermont
Infantry youngest recipient
of Medal of Honor

Edward Black, 8 ½
youngest officially
enlisted in the army

John Lincoln Clem, 12
staff sergeant, youngest
non-commissioned officer

Charley King, 12
Drum Major, Co F 49th PA
Killed at Antietam
believed the youngest fatality

Willie Lawn, 10
believed to be the youngest
wounded, lost part of his right
arm near Suffolk, Virginia

Jimmie Johnston, 6 1/2
youngest to see action aboard
the gunboat Forest Rose
son of captain John V. Johnston

WILLIE JOHNSTON, AGE 12

drummer 3rd Vermont, Medal of Honor/Peninsula Campaign

Born sometime in July 1850, William E. "Willie" Johnston, serving as volunteer drummer for regimental recruiters, joined Company D of the 3rd Vermont Infantry when his father enlisted. Due to his age, he could not formally muster in for pay, but in December of 1861 he enlisted and went along anyway serving in all the camp duties of a regular musician. As the summer campaign got underway, he was mustered in and immediately saw action in the Peninsula Campaign. The Federal Army had advanced to the outskirts of Richmond when the Confederate General, Joseph E. Johnston was wounded and replaced by Robert E. Lee. The war turned and the Union soldiers found themselves driven back toward the Chesapeake Bay. Finally, at a place called Malvern Hill, they regrouped and held their ground.

Willie, like many others, had lost most of his gear through it all, but he hadn't lost his drum, for he knew he was useless to his comrades without it. When the regiment assembled the following week for a grand review, Lincoln was there. He noticed that the makeshift band had but a few fifers and only one drum. It was Willie. The president paused to speak to the boy. He was visibly shaken to learn the boy's name was Willie. His own son, Willie, had recently died of a fever. They would meet again.

Willie became sick that summer and ended up in a hospital in Baltimore. After he recovered, he was assigned there as a nurse and orderly. In September of 1863 he was taken to Washington, DC, bathed, clipped, and fitted out with a new silk uniform. Afterwards, Willie found himself in the reception hall of the War Department facing Colonel E. D. Townsend, assistant adjutant general of the entire United States Army. The colonel spoke a few words to the crowd of dignitaries gathered there, then presented Drummer Willie Johnston, just 13 years old at the time, with the Medal of Honor. The medal had been authorized July 12, 1862, to be presented to non-commissioned officers and privates for gallantry in action. General Smith had noticed Willie's actions that summer and had submitted his name in recommendation for the medal. The picture above was taken the day of the ceremony.

President Lincoln had not forgotten the single drummer from the previous summer. Following the presentation he visited privately with Willie before the boy left to return to his duties.

Willie Johnston's story can be found in **When Johnny Went Marching** by G. Clifton Wisler, and in **Too Young to Die** by Dennis M. Keesee, along with the stories of dozens of other young boys who fought in the Civil War. Young Johnston is also the subject of a book, **Lincoln's Drummer**, written by G. Clifton Wisler

Statue of Johnston
Old Town Newhall's Veterans Memorial Plaza in Santa Clarita, California.

Edward Black, age 8 1/2

Edward Black was born in 1853, and on July 24th, 1861 he signed up with the 21st Indiana Volunteers' 1st Regiment as their drummer boy. At 8 years old, it is believed to this day that he was the youngest Union Army solider, and the youngest ever since to enlist in the armed services.

Drummer boys were in use before the Revolutionary War to maintain a "pace" or rhythm for marching or charging a line. By the time of the Civil War their use had started to decline, and in 1862 President Lincoln ordered that drummer boys be discontinued in order to stem the tide of young boys being killed in the war. The drummer boy was often out in front of the advancing troops, and the enemy often aimed at them in order to cause the advancing line to fall out of step.

Edward's Civil War drum, now in the collection of the Children's Museum of Indianapolis

Edward went home to Indiana as ordered but didn't stay for long - he re-enlisted again with his father that very same year, and they both served until the war's end. He was taken prisoner in the Battle of Baton Rouge but survived to be a Civil War veteran at the ripe old age of 11.

Unfortunately, he died at age 19 in 1872, never having fully recovered from the traumatic experience of war. He is buried in Section 16, Lot 148.

Edward is buried next to his twin brother who died less than one year after birth. Both gravestones sit next to an obelisk upon which is inscribed "Drummer Boy of the 21st Ind. Regt. at the age of 8 Yrs. 6M. The Youngest Soldier of the Rebellion." Interestingly, Edward's stone is in much better condition than his brother Edwin's, and the dates of death for Edward on the obelisk and the headstone do not match - one states 1871 and the other 1872.

Source: "Lost Indiana".

During the Civil War, many regimental bands used boys as drummers because that meant that more men were available to fight. As far as we know, Edward Black of Indianapolis was the youngest drummer boy in the war. He enlisted on July 24, 1861 in the 1st Regiment, 21st Indiana Volunteers. He had celebrated his eighth birthday just two months earlier.

JOHNNY CLEM: THE CIVIL WAR DRUMMER BOY OF CHICKAMAUGA, AGE 12

John Joseph Clem was born August 13, 1851. He was christened at St. Francis DeSales Church in Newark, Ohio where he grew up. He and his younger siblings sold produce from their parents' small farm, carrying it in a wagon. His mother was killed crossing the railroad tracks, and when John's father remarried, John did not get along with his stepmother. [Schmidt] He attended school in Newark where a primary school is now named after him. We do not know just why he felt so strongly about the War. Many at the time did. Generally preservation of the Union was their primary motivation, but a smaller number were motivated by the ideal of abolition of slavery. One factor was probably the problems with his stepmother made life at home unpleasant. His relatives joining up was probably another factor.

Johnny began cutting classes so he could drill as a drummer boy with a local unit--Company H, 3rd Ohio Volunteer Infantry. Many of John's relatives had enlisted in the army, and he felt such admiration for the president that he changed his middle name to Lincoln. Recruiters turned him down because of his age. Several regiments passed through Newark on the way to the War. Each time he tried to join he was turned down because of his age. He traveled by train to try to enlist, but he was always recognized by a friend or relative and returned home.

* * *

In his Historical Collections of Ohio, Henry Howe tells us the story of young Johnny Clem, widely known as the drummer boy of Shiloh and quite possibly the youngest to bear arms in the American Civil War. As part of his research, Howe was able to interview Clem's family in Newark, Ohio. Lizzie Clem, who was 7 years old when her older brother left home for the Army recalls the following events from the day prior to Johnny's departure:

It being Sunday, May 24, 1961, and the great rebellion in progress. Johnnie said at dinner table: "Father, I'd like mighty well to be a drummer boy. Can't I go into the Union Army?" "Tut, what nonsense, boy! you are not ten years old." Yet when he had disappeared it is strange we had no thoughts that he had gone into the service.

When dinner was over Johnnie took charge of us, I being seven years old and our brother, Lewis, five years, and we started for the Francis de Sales Sunday-school. As it was early, he left us at the church door, saying, "I will go and take a swim and be back in time." He was a fine swimmer. That was the last we saw of him for two years.

The distress of our father and step-mother at Johnnie's disappearance was beyond measure. Our own mother had met with a shocking death the year before: had been run over by a yard engine as she was crossing the track to avoid another train. No own mother could be more kind to us than our step-mother. Father, thinking Johnnie must have been drowned, had the water drawn from the head of the canal. Mother traveled hither and yon to find him. It was all in vain. Several weeks elapsed when we heard of him as having been in Mount Vernon; and then for two years nothing more was heard and we mourned him as dead, not even dreaming that he could be in the army, he was so very small, nothing but a child.

*   *   *

Separating fact from myth is always hard when it comes to celebrity soldiers like Clem. He was born Aug. 13, 1851, in Newark, Ohio, and was just 9 years old when he "ran off to join Mr. Lincoln's army" (as the song in the movie goes). He was reportedly rejected by the commander of the 3rd Ohio, who said he "wasn't enlisting infants," before finally managing to hook up with the 22nd Michigan Infantry in the late summer of 1862. The 4-foot-tall youngster made himself indispensible around camp and the following spring was officially mustered into the regiment as a musician. Intelligent despite his limited education, Clem was given the important duty of regimental marker, carrying the guidon that a unit formed its line on.

A few months later, the 22nd Michigan was heavily engaged at the Battle of Chickamauga. On Sept. 20, 1863, in the midst of a retreat, Clem found himself face to face with a Rebel colonel on horseback. As the story goes, the officer yelled, "Stop, you little Yankee devil!" Clem refused to surrender. As he later described it, he picked up a discarded rifle, pointed it at the officer — and to both combatants' great surprise, shot him out of the saddle. As one typically overheated newspaper account of the incident put it, "The proud Colonel tumbled dead from his horse, his lips fresh stained with the syllable of vile reproach he had flung upon a mother's grave in the hearing of her child."

Civil War Drummer Boy Johnny Clem Takes Arms

In a previous post we learned of the circumstances under which young Johnny Clem left home in order to enlist and fight for the cause. What follows here is an account of the manner in which the young Clem showed his mettle at the Battle of Chickamauga. The account is provided by Benjamin F. Taylor, military correspondent for the Chicago Evening Journal. Taylor writes:

You remember the story of little Johnny Clem, the atom of a drummer-boy, "aged ten," who strayed away from Newark, Ohio, and the first we know of him, though small enough to live in a drum, was beating the long roll for the 22d Michigan. At Chickamauga, he filled the office of a "marker," carrying the guidon whereby they form the lines, a duty having its counterpart in the surveyor's more peaceful calling in the flagman who flutters the red signal along the metes and bounds. On the Sunday of the battle, the little fellow's occupation gone, he picked up a gun that had slipped from some dying hand, provided himself with ammunition, and began putting in the periods quite on his own account, blazing away close to the ground, like a fire-fly in the grass. Late in the waning day, the waif left almost alone in the whirl of the battle, one of Longstreet's Colonels dashed up, and, looking down at him, ordered him to surrender: "Surrender!" he shouted, "you little d--d son of a -----!" The words were hardly out of the officer's mouth, when Johnny brought his piece to "order arms," and as his hand slipped down to the hammer he pressed it back, swung up the gun to the position of "charge bayonet," and as the officer raised his sabre to strike the piece aside, the glancing barrel lifted into range, and the proud Colonel tumbled dead from his horse, his lips fresh stained with the syllable of reproach he had hurled at the child. A few swift moments ticked off by musket shots, and the tiny gunner was swept up at a swoop and borne away a prisoner. Soldiers, bigger but not better, were taken with him, only to be washed back again by a surge of Federal troopers, and the prisoner of thirty minutes was again John Clem "of ours," and General Rosecrans made him a Sergeant, and the stripes of rank covered him all over like a mouse in a harness, and the daughter of Mr. Secretary Chase presented him a silver medal appropriately inscribed, which he worthily wears, a royal order of honor, upon his left breast, and all men conspire to spoil him, but, since few ladies can get at him here, perhaps he may be saved. Think of a sixty-three pound Sergeant, fancy a handful of a hero, and then read the "Arabian Nights" and believe them.

*   *   *

It evolved that the colonel wasn't killed, and that Clem had not shot him with a custom-fitted miniature musket, as widely reported. But the boy did evade capture by rolling himself in a blanket before finally making it back to his decimated regiment. Word of the exploits of 12-year-old Clem spread quickly among the demoralized troops.

The youngster's admirers in the press and the army didn't quibble over all the details of his heroism. Before he knew it, "The Drummer Boy of Chickamauga" was a celebrity, written up in national publications, posing for photographs, and accepting the gift of a pony. According to some sources, the song, "When Johnny Comes Marching Home Again," was based on Clem, who was promoted to sergeant. Countless youths were motivated by his example. One general, employing his best once-upon-a-time style, was moved to write his own young son:

"What shall I write to you about? I will tell you a story of a little boy who once lived in Michigan. His name is John Clem … . He was a good boy — always obeyed his Captain and always tried to do his duty like a man. Being a good boy, everyone liked him, because good boys always have a great many friends — he had many. Last summer his drum was broken by some accident and poor Johnny often cried because he had no drum to beat, but he always kept up with his Company in either hot or cold weather and often he had to sleep on the cold damp ground without a blanket … . Johnny will make a great man some of these days and so will any boy who is obedient and faithful in the performance of his duty."

*   *   *

Johnny in October 1863, was captured in Georgia by Confederate cavalry while detailed as a train guard. The Confederate soldiers took his uniform away from him which reportedly upset him terribly--especially his cap which he said had three bullet holes in it. [Bennett] He was exchanged a short time later, but the Confederate newspapers used his age and celebrity status to show "what sore straits the Yankees are driven, when they have to send their babies out to fight us." Another account indicates that he was held somewhat longer. According to one source, family members complained that he looked like a "dreadful corpse" and was said that when he came home he "couldn't have weighed over 60 pounds." [Bennett]

*   *   *

Clem didn't quite approach "great man" status as he grew older, but he was able to seize advantage of the connections fame had brought him. After his skimpy education torpedoed his attempt to enter West Point, he prevailed upon President Ulysses S. Grant to appoint him a second lieutenant in the Regular army. Clem served from 1871 to 1916, when he retired as a mildly competent but beloved major general. He was the last Civil War veteran to leave active duty. He died at his Texas home on May 13, 1937, and was buried at Arlington National Cemetery.

Curiously, for all of his adult life, Clem often was erroneously referred to as "The Drummer Boy of Shiloh." This was no fault of Clem's. Rather, it was the result of an 1871 newspaper article that mistakenly identified him as the youngster who'd famously had his drum destroyed by a shell at the Battle of Shiloh in April 1862. That incident — most likely apocryphal — inspired one of the most popular songs of the war, "The Drummer Boy of Shiloh," which in turn begat a play by the same name. Performances of this patriotic tearjerker were a staple of school fundraisers and veterans' gatherings well into the 20th century and kept the name alive in the public mind.

The story was widely circulated as a pamphlet and found its way into Clem's service jacket. The appellation stuck, although a number of other young soldiers would always claim to be the real "Johnny Shiloh." Writers and historians used the 1871 article for more than a century without bothering to check its claims against Clem's service records. If they had, they would have realized that Clem's participation at Shiloh would have been impossible. At the time, the 22nd Michigan hadn't even been organized, and in any case Clem had yet to join the regiment.

Clem never really claimed to be Johnny Shiloh. He was satisfied being, as his gravestone at Arlington reads, "The Drummer Boy of Chickamauga."

Sources

Bennett, Kevin. "Clem: Newark's most famous veteran," The Advocate (Newark), June 17, 2002.

Casamer, Douglas M. The Michigan 22nd Infantry and the Men Who Served.

Casamer, Douglas M. E-mail message, March 9, 2004.

Clem, John L. "From Nursery to Battlefield," Outlook" magazine CVII (1914), 546-547.

Schmidt, Barb. E-mail message, February 11, 2011. Schmidt is a distant relative and her family has been collecting information on Johnny.

Taylor, Nicholas J. E-mail message.

Wiley, Bell Irvin. The Life of Billy Yank: The Common Soldier of the Union (Louisiana State University Press, 1952. (The edition quoyed here is the 1978 reissue.)

Charles "Charley" King, Age 12

12 Years 5 Mo. and 9 Days. Old when he enlisted.

Great sacrifices were made at the Battle of Antietam. And one very small one.

Charles 'Charley' King was from West Chester, Pennsylvania. He was the oldest child in his family, but only 12 years old when the Civil War began and northern units began to organize and train. Charley was fascinated by all the activity and he also loved to make music. He begged his father to allow him to enlist as a drummer boy. Attitudes toward children were different at the time, but even for Civil War parents, 12-years old was very young to enlist in the Army. Boys that age, however, can be very persistent. And Charley managed to gain an ally. Company F Captain Benjamin Sweeney noticed him practicing his drumming near the camp where the 49th Pennsylvania was training. Sweeney was not only involved with the training, but also recruiting soldiers. Sweeney was impressed with both Charley's determination and drumming skills. He had a talk with Charley's father that drummer boys were non-combatants and not really in danger. He said the boys were usually behind the lines and in safe positions helping with the wounded rather than on the battlefield. He promised he would keep Charley safely away from danger and look after him. Of course there was no way that he could possibly ensure this. But in fairness to captain Sweeney, very few people at the time realized what a blood bath the Civil War would become. Most had highly unrealistic, romantic ideas about battle and thought the War would soon be over. So Charley's father gave his permission. We are not sure if mother had any say in this. The family agreed to let the oldest of their five children enter the war.

VR 12.31.1861

Young Patriotism.—Charles King, son of Pennell King, of West Chester, not quite 13 years of age, is drum-major of the 49th Regiment, now encamped about ten miles from Washington. We doubt very much, if in the whole army in the field, there is another so young possessing the rank of drum-major. When the three months men started out he played for a company as far as Harrisburg, but being so young, his parents would not permit him to accompany them further.— When the next call was made for men, he accompanied Capt. Sweney's company to Harrisburg. While there he insisted so strongly on following the men to the battle field, that Capt. Sweney interceded, and with the promise that he would be well guarded, the parents' consent was obtained. He was then made drummer of the company; soon after he was taken as drummer in the band of the Regiment, and now he has been made drum-major of the Regiment. He was so taken with going that his father would very frequently at nights, find him setting up in bed "marking time" on the head-board of his bed. Another good trait of his, which might be an example for older ones, is that whenever he draws his pay, the money is sent home, where it is deposited for him.

Village Record, December 31, 1861

Charley performed well as a musician-drummer boy with the company, impressing the men and officers so much with his drumming that he was promoted drum major of the field music of the Forty-Ninth Pennsylvania Regiment. It was a distinct honor for such a young volunteer. After participating in the 1862 Peninsula Campaign, Charley was a veteran, something that few of his age could claim.

With the movement of Lee's Confederate Army of Northern Virginia into Maryland in September of 1862, Charley marched with the rest of the Federal Army toward Western Maryland and a showdown with the enemy on Sept. 17, 1862, The Battle of Antietam - the bloodiest single day of fighting in the Civil War.

The 49th Pennsylvania was in a reserve position. Confederate artillery, however, hammered the Regiment, during the battle, wounding several men. Enemy artillery exploded near the Forty-Ninth Pennsylvania, wounding several men including Charley, who was shot "through the body" by a piece of shrapnel. He fell into the arms of H.H. Bowles of the 6th Maine. Bowles carried Charley to a field hospital in the rear. Medical care was, however, primitive at the time, and Charley died 3 days later from his wound. He was one of the nearly 25,000 men killed and wounded at Antietem. Charley is believed to be the youngest soldier of either army to be killed during the 4 years of fighting. A few younger boys served in the war, but Charley apparently was the youngest combatant killed as a result of a battlefield wound.

Charley's father was informed and retrieved his body after the battle and laid him to rest near his home in West Chester, at Old Cheyney Cemetery.

VR 10.7.1862

On the 19th ult., from wounds received at the battle of Antietam, CHARLES EDWIN, son of Pennell and Adaline King, aged 13 years, 5 months and 15 days. *10.7.62*

A little more than one year ago Charlie left his home, to go forth to the call of his Country in the capacity of drummer, under Captain Sweeney, of Company F, 49th Regiment, P. V. Daily are we called upon to record the names of brave men, who have fallen a sacrifice on their country's altar, that gives us pain, but how much keener is the anguish, when the tidings reach us that a child of such tender years has been made a victim by some rebellious hand. He was a true little hero. In the wearying marches by night and by day, he never murmered or complained. His mind was beyond his years. By his good behavior and gentlemanly deportment, he won many friends who deeply deplore his irreparable loss. But their loss is his eternal gain. His sorrow stricken and bereaved parents, though the stroke falls heavily upon them, have the consolation to know that their child has done his duty and died nobly. He is no longer a drummer boy, weary and footsore upon the field of battle, but rests now at the feet of Jesus singing hymns of praise, and awaits the coming of his parents and brothers and sisters, from whom he will never be parted again, but will dwell together forever and ever.

Village Record, October 2, 1862

*   *   *

Information for this section was contributed by Andy Waskie.
Image Courtesy of The State Museum of Pennsylvania

Resources & Reference Material
Secondary Sources

* Antietam: The Photographic Legacy of America's Bloodiest Day, by William A. Frassanito, 1978.
* History of the 49th Pennsylvania Volunteers, by Robert S. Westbrook, 1898.
* Pennsylvania at Antietam: Report of the Antietam Battlefield, by Antietam Battlefield Memorial Commission, 1906.
* History of the Pennsylvania Volunteers: 1861–1865, by Samuel P. Bates.

Charlie King Memorial

On Saturday, June 18, 2005, Brendan Lyons, of West Chester, PA, completed his Eagle Scout project, culminating in a memorial service for Charley King and recognizing his ultimate sacrifice as the youngest Union soldier in the Civil War to be killed as a result of battle wounds.

A memorial stone was placed in West Chester's Greenmount Cemetery, where both Charley's parents were buried.

Drums of War

By Edith Morris Hemingway and Jacqueline Cosgrove Shields

A special edition of Broken Drum (under new title and cover art), licensed and edited by Scholastic specifically for elementary schools to be sold in Scholastic Book Fairs and Book Clubs nationwide!

About the book:

Charley King, a twelve-year-old boy from Pennsylvania, is caught up in the excitement and patriotism of the Civil War. Proud and eager to serve his country, Charley volunteers as a drummer boy, beginning the perilous journey that takes him from the defense of Washington to the horrific battle of Antietam. Based on an actual person, Drums of War is a novel that will transport you to the front lines of the War Between the States, where you will march alongside Charley right up to the stunning, gut-wrenching conclusion.

Charles E. King,
12 years old when he enlisted in the PA 49th Volunteers
A Tribute To Charley King

Jackie Shields and I were invited as guests and speakers during the ceremony. Following is my speech given in honor of Charley King:

June 18, 2005
West Chester, Pennsylvania

Good afternoon. It is an honor for me to be included in this memorial service and tribute to Charles King. Over the course of writing Broken Drum, which, incidentally, took us 5 years, we became very attached to Charley, as if he were an integral part of our own families.

Charley's faded photograph, on display at the Antietam Battlefield Museum, was our inspiration for writing Broken Drum. We knew there was a story behind that solemn-faced, most likely nervous, young boy, who was a mere 12 years, 5 months, and 9 days old on the day he enlisted in the Pennsylvania 49th Volunteers. When I counted back the days, months, and years to his birth date, I found that Charley was born in April of 1849, exactly one hundred years earlier than my own brother's birthday. A connection was made.

As mothers ourselves, Jackie and I could not imagine allowing our sons to go off to war at 12 years old. It would be difficult at any age, as many mothers know right now. Our first thoughts were that Charley was either an orphan or a runaway, and that's when our imaginations took off. Our research brought us to Charley's origins here in West Chester, Pennsylvania.

Broken Drum is not a biography of Charley, by any means. Because of limited information, we did fictionalize a number of the events, but we also thoroughly researched the Pennsylvania 49th, as well as the everyday lives of drummer boys in the Civil War. We hope that we realistically portrayed the tremendous responsibilities that drummer boys shouldered during our Civil War and that we acknowledged the significant contributions they made to their regiments and to their country.

In the course of bringing our fictional Charley to life for our readers, we also hope that we portrayed the true character of Charley King, who was a favorite among the soldiers of Company F, and, I am most certain, a well-loved member of his family. As Jackie and I dragged out the writing of the book, my husband teased me more than once that we were taking so long because we didn't want to kill Charley. Well, that's true, we didn't want Charley to die, but we had no choice as we could not change history.

Jackie and I did most of our writing at my home on top of Braddock Mountain in Maryland, which looked out across the valley to the long impressive ridge of South Mountain. To this day, I do not look at that ridge without thinking of Charley's trek over that mountain and the skirmish he was involved in at Crampton's Gap the day before he went into his final battle at Antietam. I have been humbled by the more than 23,000 candles illuminated at Antietam Battlefield on a dark night each December, knowing that Charley was one of those 23,000 wounded and killed on that horrendous day—each one a hero in his own right.

Brendan Lyons, thank you for remembering Charley King and for bringing this memorial service to fruition. Andy Lefko, thank you for searching us out and inviting us to participate. We are honored.

~Drop me a line at: edie@ediehemingway.com

WILLIE LAWN, AGE 10

Willie Lawn. "Soldier

Ten-year-old Willie Lawn was wounded near Suffolk, Virginia, April 23, 1863. He lost part of his right arm.

Jimmie Johnston, age 6 1/2

"Powder Boy" James V. Johnston

Jimmy Johnson-Powder Monkey This uniform was presented to James V. "Jimmie" Johnston by the crew of the U.S. gunboat **Forest Rose,** for gallantry in action during the Civil War. He had accompanied his mother to visit his father, Captain John V. Johnston, aboard the gunboat in peaceful waters on the Mississippi. As the vessel approached Waterproof,

Louisiana, in February 1864 it came under attack by a Confederate force. When the gunboat's regular powder monkey, who carried powder to the gunners, was killed early in the battle, young Jimmie took his place until the Confederates were repelled. The crew presented this uniform to the six-and-a-half-year-old boy they called "Admiral Johnston" for his bravery

* James "Jimmie" Vincent Johnston was born September 23, 1857, and "was the youngest person who rendered effective service in battle" during the Civil War at age 6 ½.

* His father, Captain John V. Johnston, was executive officer on Admiral Andrew Hull Foote's flagship during the attack on Fort Donelson.

* On February 13, 1864, his wife and son Jimmie were aboard the Forest Rose gunboat with him when they were unexpectedly attacked by a force of 5,000 Confederates who were trying to cross the Mississippi to reinforce General Joseph E. Johnston, who was confronting General Sherman in Georgia.

* From February 13 to 15, the Forest Rose resisted their attempts to cross the river. Captain Johnston attempted to keep his son below deck with his wife, but the boy escaped to be with the gunners several times.

* Finally, when the regular powder boy was shot and killed, Jimmie took on his role.

* When his father discovered him, he asked where he had gotten the load of powder. Jimmie replied, "Why, Tommy [the powder boy] had his head shotted off over there, an' I'm a-carrying' his powder."

* Thanks to little Jimmie's help, the Forest Rose repelled the Confederate force successfully.

* After the battle, in appreciation of the boy's heroics the sailors nicknamed him "Admiral Jimmie" and made him a miniature sailor's uniform.

http://www.civilwarmo.org/educators/resources/info-sheets/jimmie-johnston-child-soldier-civil-war
http://www.dailymail.co.uk/news/article-2237476/Civil-War-era-photograph-collection-displays-dignity-young-men-join-battle.html
http://www.old-picture.com/1860s-index-001.htm
http://www.civilwarmo.org/educators/resources/topics/soldiers-experience

Books of interest, suggested reading:

A History of Missouri 1860-1875 by William E. Parrish

Children of the Civil War by Candice Ransom

Civil War for Kids by Janis Herbert

Civil War in Missouri Day by Day by Carolyn Bartels

Civil War on the Western Border by Jay Monaghan

Civil War's First Blood by James Denny, John F. Bradbury

General Sterling Price and the Confederacy by Thomas C. Reynolds

I'll Pass For Your Comrade: Women Soldiers in the Civil War by Anita
    Silvey

Kids During the American Civil War by Lisa A. Wroble

Missouri Brothers in Gray by William Jeffery Bull

Missouri Slave Narratives by Federal Writers' Project

Nobody's Boy by Jennifer Fleischner

Sterling Price: Portrait of a Southerner by Robert Shalhope

Boy Recipients of the Medal of Honor
During the American Civil War

the Medal of Honor established during the Civil War

Julian Scott, 15
Drummer, E Co., 3rd Vermont Inf.

William Horsfall, 15
Drummer, G Co., 30th Mass Inf.

Willie Johnston, 12
Drummer, D Co. 3rd Vermont Inf.

Orion Howe, 13
Musician, C Co., 55th Ill inf.

John Cook, 15
Bugler, B Battery, 4th US Artillery

JULIAN SCOTT, FIFER AND ARTIST

a self-portrait sketch of Julian Scott

a sketch by Julian of his home

Julian Scott was 15 years old when he enlisted as drummer and fifer with the 3rd Vermont Infantry. He liked to sketch pictures and frequently sketched scenes and people in camp as well as scenes from the battlefield. During the 7 Days Battle retreat from Richmond in the early summer of 1862, his unit was involved in combat with the Confederate Army which required crossing a creek. As his regiment retreated across the creek, many wounded were trapped on the Confederate side. Julian made several trips under fire across the creek to bring back wounded soldiers. For his actions that day, he was awarded the Medal of Honor.

Julian's commanding officers recognized his artistic skill and were instrumental in getting him into an art school late in the war. He later became a renowned artist of Civil War art. Included herein are copies of some of his letters and of his art.

The following extracts are from a letter by Julian A. Scott, written to his father in Johnson, and dated near Kent Court House, Va., May 11:

A soldier of the 5th Regiment and I, went in advance of our army, so as to get some trophies the rebels left behind. We had got about a mile in advance when I spied a rebel scout. We resolved to take him, dead or alive; so we parted --- my comrade going one way, and I another, so as to come on both sides of him, in order to take him alive. If he shot one of us, the other would be sure to bring him down.. I had my revolver cocked, and ready to take him if he meant to get away. We caught up with him at the same time he spied my comrade. I saw the rebel take aim, and drew my revolver on him, so that if his took effect he must come too. He fired- the shot passed by. I hallooed "halt", pointing my revolver at him. He saw his fate, and threw down his gun. I walked up to him and took it, and my comrade and I took him to General Smith, feeling proud of our prize. The general spoke to me and said, " Scott, where did you get that fellow". I told him. He asked me if I had a gun. I told him I have not. He then said I might keep the one the rebel had. He asked the rebel a few questions about the secesh army. I then took him to the provost Marshal, and left him.

After the Battle of Williamsburgh he says:

I passed over the battlefield: It was covered with blood. There was one poor fellow that was dying near me. When I spoke to him, he said he was dying, and wished that his mother could know his fate. His father was killed at Bull Run: he enlisted to get money for his mother. He died in a few minutes. I closed his eyes , and tied my handkerchief around his head. To his order, he was buried the next day. There was one Lieutenant of Co. I, 5th North Carolina Regt., who was shot through the heart. The Col. Of the 49th Penn knew him, and said he would give five dollars to anyone who would bury him. Three of us took him under an old tree, and gave him a decent burial. I took his watch and drinking cup. In his wallet was his commission, looks of his wife, and his mother's hair, a gold pen, and some rebel postage stamps. His name was Samuel P. Snow.

My health is good. I take delight in fighting the rebels.

Submitted by Deanna French.
Post-War Articles
LAMOILLE NEWSDEALER: NOVEMBER 9, 1870

Julian Scott's art

More information can be found on these web links:

http://www.google.com/imgres?imgurl=http://townofjohnson.com/Portals/12/
    julianphoto.JPG&imgrefurl=http://townofjohnson.com/History/JU-
    LIANSCOTT18461901/JulianScottArtistoftheCivilWar/tabid/953/Default.
    aspx&h=555&w=405&sz=22&tbnid=iyOAeda6hRFZ8M:&tbnh=59&tbn-
    w=43&prev=/search%3Fq%3Djulian%2Bscott%27s%2Bart%26tbm%3Dis-
    ch%26tbo%3Du&zoom=1&q=julian+scott%27s+art&usg=__CdWQ_as_PqI-
    Ry1hdVd53pDCpOqg=&hl=en-US&sa=X&ei=wad2UoqZEOy2sATCq4H-
    QBQ&ved=0CCQQ9QEwAQ
http://vermontcivilwar.org/get.php?input=5226
http://vermontcivilwar.org/museum/moh/bios.php?input=5226

Julian Scott was a drummer and fifer
Medal of Honor
This soldier was awarded the Medal of Honor

Julian A. Scott

Rank and Organization: Drummer, Co. E, 3rd Vermont Infantry.
Place and date: Lees Mills, VA, 16 Apr 1862.
Entered service at: Johnson.
Born: 15 Feb 1846, Johnson.
Died: 4 Jul 1901.
Buried: Hillside Cemetery, Scotch Plains, NJ
Date of Issue: Feb 1865.
Citation: Crossed the creek under a terrific fire of musketry several times to assist in bringing off the wounded.

JULIAN SCOTT

Lamoille County has reason to be justly proud of its son, this very promising young artist. He is a native of Johnson, and a son of C.W. Scott, for many years a watchmaker and jeweler there. We give below some notice of him and his work from the papers of the state.

JULIAN SCOTT'S DRUMMER BOY. --- When the tocsin of war sounded a Vermont boy left his country home and joined the ranks. To relieve the weariness of camp life he resorted to sketching and painting, for which he has always possessed a liking, but he had been in circumstances not favorable to art. After the war he turned his recreation into a profession, and after only so brief a career, gives unmistakable proof of have been born an artist, which means much. His studio is something of an arsenal, for he has accoutrements enough about it to equip a small force. He is writing a history of the war on canvas --- a history of which himself is a part, as all soldiers are; pictures photographed on his brain with fire and sword, the valor of soldiery, the boom of cannon, the anguish of dying and the awful heart- sickening faces of the dead.

He has a cartoon of the "Battle of White Oaks Swamp" at the moment when General Smith and his staff came forward --- and by their presence and efforts rallied the forces and prevented defeat, which he will produce in oil next autumn.

His "DYING DRUMMER BOY" should have been placed on exhibition at the Academy --- that is, if merit implies "Oughtness." Only the artist's "perversity" prevented. It is such a sad, sad picture! Enough to bring tears from the hardest eyes, even eyes hardened by such scenes---scenes of lone, agonizing dying, with the face whitening at every moment; of the swift pictures of home fleeing forever; the last hunger for the dear ones far away gnawing at the faint heart, the dream of ambition over, and glory and pomp of battle fading out into the awful shadow of forgetfulness, and all this on the "Dying Drummer Boy's Face."

With one so young, so gifted, so genial as Scott, one cannot but leave the heart's benediction

M.A.E.W

So wrote an art loving correspondent of the New York Times, a year ago. We hear that Mr. Scott, yielding to the expressed desire of many members of the Legislature to see some specimens of his skill, has sent for this painting, and it will be at Montpelier this week. It will not be on public exhibition; but will be hung for a few days, probably in some room at the Pavilion or at the Capital, where it can be readily seen by any who care to see it.

Julian Scott, whose name and fame are chronicled elsewhere, is at Montpelier soliciting from the legislature an order for a historical painting representing some scene in the experience of the "Old Brigade," of which he was a member. All he asks is pay for his time, trusting to the reputation the work will give him for any further satisfaction. He is willing to spend two years upon the picture upon these terms. The offer is certainly a very liberal one and we hope the Legislature will not hesitate to accept it. It is time our State House began to receive some such adornment as is suggested.

WOODSTOCK STANDARD.

LET VERMONT ENCOURAGE VERMONTERS: --- We trust the State will adopt some wise policy about the matter and not leave the decoration of the Capital to chance or haphazard. A little help, at the right time, while they are alive and in need of work and assistance, would be of greater benefit to the State than to them. We want living art and not dead art. It should be the object of the State to be the patron of a living artist, and illustrate its history by their works.

MONTPELIER JOURNAL.

Submitted by Deanna French.
Obituaries

Washington Post

5 Jul 1901

New York, July 4 - Col. Julian Scott, the well known artist, is dead at his home in Plainfield, N.J. Col. Scott was born at Johnson, Vt., February 15, 1846. When the civil war broke out he enlisted in the Third Vermont Regiment as a musician. Later he was appointed on the staff of Gen. "Baldy" Smith. He was the first man to receive a medal of honor for official bravery on the battlefield. This was voted to him during Secretary Stanton's term of office. At the close of the war Col. Scott entered the Academy of Design in New York and finished his studies in Paris.

Contributed by Erik Hinckley.
The Daily Press, Plainfield, NJ, 8 July 1901

DEATH HAS CLAIMED COL. SCOTT

The well known artist passed away at Muhlenberg Hospital yesterday morning. HAD A REMARKABLE CAREER:
HIS PAINTINGS ARE TO BE FOUND IN MANY PLACES:

WAS ONE OF PLAINFIELDS MOST PROMINENT CITIZENS: HAD BEEN ILL FOR SOMETIME. HIS FAMILY AT HIS BEDSIDE

About His Life

In the death of Julian Scott, which occurred yesterday morning about 5 o'clock at Muhlenberg Hospital, Plainfield has lost of its most prominent citizens, and the country in general has lost one of her best artists. He had been ill for some time. A few weeks ago his condition was such that it was thought best to take him to the hospital. With the best of treatment he did not improve, but gradually grew weaker and finally he just passed away as though in sleep.

Just before he died he seemed to be more conscious than at any time recently, and opening his eyes, he seemed for a moment as though possessed of greater strength. It was thus that he died. Mrs. Scott and her daughter recently arrived from Paris, and Col. Scott's brother, H. P. Scott of Kansas City were with the Colonel when the end came.

With the death of Col. Scott a person of more than usual importance is taken away. At an early age he figured largely in matters of note. Born in Johnson, Lamoille County, Vermont, Feb. 15, 1846, the son of Mr. & Mrs. Charles Scott, he had a history that brought him the highest recognition, from all points of the globe. His mother was an artist, and his father was a genius. This happy combination resulted in bringing renown to the son. In his boy hood days he attended the academy at Johnson, where Admiral Dewey was a student.

At the age of 15 he was filled with the spirit of patriotism. It being the time of the outbreak of the Civil War, he enlisted, as a musician, in Co. E. 3rd Vermont volunteers. He served two and a half years with honor. Before he completed his service he was appointed on the staff of Gen. Baldy Smith. For acts of special bravery in wading a stream and rescuing wounded Union soldiers, he was the first one to receive the Medal of Honor. This was voted to him by Congress during Secretary Stanton's term of office.

During the service his talent as an artist developed rapidly and he made a number of war sketches which afterwards became famous. He had a natural talent for drawing and painting, and after the close of the war he entered the Academy of Design in New York, where he remained for some time. He afterwards went to Paris and completed his studies. Upon returning to this country he opened a studio in New York, and remained there for a few years.

It was 26 years ago that he came to Plainfield and opened a studio on West Point Street. He made many friends by his genial and affable manner, while through his work he gained a wide reputation. In 1888 and 1889 he was in Arizona and New Mexico, having been a special commissioner by President Harrison to inquire into the condition of the Navajo and Moqui tribes of Indians. He became very friendly with the Indians, and they elected him to membership to one of their secret societies. He also succeeded in working himself through a Masonic Lodge of Indians. His report to the government was one of the most complete ever made, and was accompanied by some forty pictures, all of which were compiled with the eleventh census.

Col. Scott was a member of many organizations of prominence, including Jerusalem Lodge No.2 6 F&AM of this city, The Medal of Honor Legion, of Washington, Clover Club of Philadelphia, Sons of The American Revolution, and The National Academy of Art and Artists Fund Society, of New York.

His paintings are to be found in most every part of the country, and among them are some of rare value: "The Battle of Antietam" is in the 7th Reg. Armory, New York, it having been presented to the regiment by the late Elliot F. Shepard. "The Rear Guard at White Oak Swamps" is the name of a valuable painting in the State House, at Montpelier, Vermont, it having been purchased out of an appropriation made by the State. "The Death of General Sedgwick" which hangs in the Plainfield Art Gallery, is the admiration of all who have seen it. The State of Connecticut has of late been trying to secure the picture to be placed in its State House, but as of yet, all efforts have been unsuccessful.

Colonel Scott was one of a few who were instrumental in establishing Plainfield's Art Gallery, and he always took much interest in it. There are also paintings by Mr. Scott in the Union League Club and the Metropolitan Museum of Art in New York. The late Mrs. Hemmingway, of Boston, who founded an art gallery and museum of natural history, in that city purchased many paintings from Mr. Scott. Notable among these is "The Song of the Ancient People," depicted in eleven water color paintings.

Mr. Scott worked steadily up to the time of his illness, having plenty to do. He courted retirement on account of this fact. At times, he was a familiar figure about the city and he always had a kind word for those he knew. He was a member of the Protestant Episcopal Church, and at times attended Grace Church in this city.

The deceased possessed many relics, pictures and trinkets of rare value. A great many Indian relics were in his studio. He also had one of two original death masks of Napoleon which experts say are original. He came into possession of it through some old junk owned by a local dealer, and purchased it for a small price. Throughout this city and this state there are many of his paintings in private collections, and have a rare value. He received his title from Drake's Zouaves of New Jersey, of which organization he was a member.

Col. Scott leaves a widow and one daughter, who have been in Paris for eighteen years, where Miss Scott is an artist, has been completing her art studies. He also leaves a brother, H. P. Scott, a lawyer in Kansas City, and two sisters, Mrs. Ladd-Davis of Brooklyn, and Mrs. Z. L. Carpenter of Kansas City.

The funeral services will be held from Grace P. E. Church Sunday afternoon at 4o'clock, and the members of Jerusalem Lodge, No. 26. F&AM will have charge. The members will meet at the Lodge Rooms at 3 o'clock. The members of Anchor Lodge No. 149(?) are respectfully invited to attend.

Submitted by Deanna French.

See also Scott's obituary in the New York Times
Funeral

PLAINFIELD COURIER: NEWS, PAGE 5

JULY 5, 1901

BURIED WITH DUE HONORS

Funeral for Colonel Julian Scott at Grace Church

The funeral was in charge of Free Masons and concluded with a touching tribute from the G.A.R. With the impressive Masonic ceremony and the honors of war, in the presence of a large number of friends, the late Julian Scott, the well known artist, was laid to rest yesterday afternoon on the highest point and one of the prettiest parts of Hillside cemetery. Representatives from many organizations to which he belonged, military men, business men, and private citizens, gathered to pay their respects to their late associate, and to take a last leave of him.

Grace Episcopal Church, where the service was held, was filled with the large company of those who had known Colonel Scott socially, professionally, and in a business way during his many years residence in this city, who had some pleasant recollections of him, or had been attracted within the circle of his acquaintance, by some one or more of the many attributes of his genial, sympathetic, unassuming nature. The solemnity of the rich Episcopal ceremony pronounced by Rev. E.M. Rodman, a Masonic Brother, together with the choral service, was very impressive. The funeral was in charge of Jerusalem Lodge, Free and Accepted Masons, and the members of Anchor Lodge were invited to attend.

About fifty of the Masons under the direction of Marshal, Robert A. Meeker, assembling at the asylum at three o'clock, marched in a body, with the insignia of the order draped, to escort the remains to the church and to perform the last rites at the grave. Beside the hearse walked the pall bearers, Dr. D.C. Adams, William H. Sebring, J. Hervey Doane, Harry W. Marshall, Dr. M.S. Simpson, and Calvin Rugg.

The body reposed in a handsome casket over which was draped the American Flag, the emblem that had inspired the ardor and enthusiasm of the youth who left home to take his first steps in the great world, apart from the influence of home, when his country was at strife with herself in the Civil War. Kindly hands had added to this emblem tender reminders of warm friendships, in the wreath of clinging ivy and fragrant from the Veteran Zouave Association of Elizabeth and other thrones friends, and the pillow of carnations and roses from the Free Masons, and the white lambskin apron, one of the emblems of the order.

At the church the procession was met by the surpliced choir and Rev. E.M. Rodman, who read the form for burial of the dead. The selections by the choir under the direction of the organist, Arthur Freeman, were, processional, "Softly now the light of day. a funeral chant, Peace perfect peace, and "Ten thousand times ten thousand.

At the grave the Masonic prayer was pronounced by Rev. George Hauser, and the burial service by Councilman, B. Frank Coriell, concluding with a short discourse and benediction by Rev. Mr. Hauser. As the brethren formed a circle and lowered the casket, the brilliant rays of sun, breaking through a rift of clouds, lighted the grave and its lining of evergreen boughs, and silhouetted the surrounding company against the clear sky. The Masons casting sprigs of cassia, the emblem of immortality, into the grave, retired. Then a detail from Winfield Scott Post G.A.R., approached, under command of Col. Albert G. Perry, with a firing squad, composed of James Baglin, John R. Vail, Edward Vanderweg, Israel Compton, John Finley, Alex Sargent, Edward DeVine, and Preston Goodfellow. Three volleys were fired in salute to the dead, and as the echoes died away, far in the distance were heard a lone bugler sounding "Taps". It was a feeling and deeply impressive leave-taking of a soldier, and many in which he, whom they came to bury, had participated on the field of battle.

Submitted by Deanna French.

WILLIAM HORSFALL, A DRUMMER BOY FROM NEWPORT WHO BECAME A CIVIL WAR HERO AT THE AGE OF 15

This story was written by reporter Cindy Schroeder. She can be reached at cschroeder@nky.com

William Horsfall was just 14 when the Newport youth and three friends stowed away on a steamship bound for a Federal regiment in the early days of the Civil War.

At the last minute, Horsfall's companions had a change of heart. They ran ashore as the steamer prepared to leave the Cincinnati Wharf.

Horsfall continued his journey, however, earning the young stowaway a place in history. Months later, at age 15, the little drummer boy saved the life of his wounded commanding officer when he was trapped between the lines on a Corinth, Miss., battlefield.

After the war, Horsfall commanded William Nelson Post GAR of Newport, and he published several songs and war poems. But his teen-aged act of heroism remained his claim to fame.

In his personal account from the Campbell County Historical Society archives, Horsfall described how he "left home without money or a warning to (his) parents" and "stealthily boarded the steamer, Annie Laurie, moored at the Cincinnati Wharf at Newport (on) the 20th of December 1861."

The stowaway stayed hidden "until the boat was well under way." When Horsfall was discovered, he said he was an orphan, and was allowed to remain on board.

On Jan. 1, 1862, Horsfall enlisted as a drummer boy in Company G, First Regiment Kentucky Volunteers, at Camp Cox, Va. (now Charleston, W.Va.).

At the time of his enlistment, the Newport teen stood 4 foot 3 and listed his occupation as "schoolboy."

During the Siege of Corinth on May 21, 1862, Horsfall, who then described himself as "an independent sharpshooter," recounted how a Union captain was wounded in "a desperate charge across (a) ravine," and left between the lines.

"Lt. Hocke ... said, 'Horsfall, Captain Williamson is in a serious predicament. Rescue him, if possible.' So I placed my gun against a tree and, in a stooping run, gained his side and dragged him to the stretcher bearers, who took him to the rear."

The drummer boy continued to take part in marches with his regiment for the rest of 1862. During the charge at Stones River near Murfreesboro, Tenn., Horsfall was surrounded by hostile infantry, but the Rebels "took pity on his youth," enabling him "to run for his life," he wrote.

About a year after enlisting, Horsfall became so ill that he had to be hospitalized. Twice in the next three years, he re-enlisted, before receiving a $400 bounty upon his final discharge in 1866.

After the war, Horsfall tried to join a volunteer fire department in his native Northern Kentucky, "but he was turned away because he was too young," said Bill Bright, curator with the Kentucky Historical Society.

By age 46, Horsfall's rheumatism and heart disease had become so bad that he required a live-in caretaker.

Horsfall received his first Medal of Honor on Aug. 17, 1895. A second medal was presented for the same citation in 1904 when the design was changed.

The once-young hero died at age 75 in 1922. Today, a rare historical marker stands near his grave in Evergreen Cemetery in Southgate.

"I think his youth is what sets him apart more then anything," said Bill Bright, curator with the Northern Kentucky Historical Society.

"A lot of people are fascinated by the story of the little drummer boy who, somewhere along the line, picked up a rifle, and, for all intents and purposes, became a private in the infantry, saving an officer's life."

Today, Horsfall is recognized as one of the youngest recipients of the Congressional Medal of Honor. The award was created during the Civil War to recognize members of the armed forces who display acts of heroism beyond the call of duty.

WILLIE JOHNSTON, DRUMMER

drummer 3rd Vermont, Medal of Honor/Peninsula Campaign

Story in section about the youngest.

THE HOWE BROTHERS AND A MEDAL OF HONOR
LYSTON HOWE, AGE 10 ORION HOWE, AGE 13

Youngest Soldier in Union Army '61 to '65.

Born Aug 27, 1850.
Age: 10 years, 9 months and 9 days, at enlistment.

Co I - 15th Ills Vol Iny
Lyston H. Howe

The stories of Lyston and Orion Howe can be found in G. Clifton Wisler's book, *When Johnny Went Marching*. Their father, William Howe, was a veteran musician from the Mexican War. He had taught both boys to play the drum, and they were good. Lyston was still 10 years old when he followed his father into the 15th Illinois Infantry. Orion was made to stay in Chicago attending school. The father and son served in Missouri, until weather conditions caused Lyston to run a dangerously high fever. Believing him near death, he was discharged in October 1861 and sent to his grandmother in Chicago. [Their mother had died in1852; the family had moved from Portage County, Ohio to Waukegan, Illinois.] But he surprised all and recovered. William became fife major for the newly formed 55th Illinois Infantry and once again, Lyston got to go and Orion was left behind. This time he was determined to go, sneaked aboard a train, then a supply boat. Arriving in Memphis, Tennessee, Orion persuaded Lieutenant Colonel Oscar Malborg to accept him as drummer for Company C.

Orion soon became the pet of the whole outfit. From September 1862 when he enlisted at age 13, he continually amused his companions, outwitted officers, and dreamed up mischievous schemes for his brother and the other drummers. The two brothers found themselves on their own when their father left the army in February 1863. In May, the 55th became part of the Vicksburg Campaign. On the 19th, the 55th was ordered into action against fortified Confederate positions. Advancing along Graveyard Road, the regiment became trapped in a narrow ravine. As men fell all around, Orion rushed out among the fallen to retrieve their cartridge boxes Ammunition was running low. The colonel, fearing for Orion's safety, sent him and at least two others, back to the main army with a request for more ammunition.

They took off across the battlefield dodging bullets along the way. His companions were killed. Halfway to safety, Orion went down with a musket ball through his right leg. But he got up again and continued on to report to General Sherman, in spite of pain and loss of blood, the terrible situation of his unit. Sherman sent relief and ordered Orion to the hospital. An historian wrote of Howe: "We could see him nearly all the way . . . he ran through what seemed a hailstorm of canister and musket-balls, each throwing up its little puff of dust when it struck the dry hillside. Suddenly he dropped and hearts sank, but he had only tripped. Often he stumbled, sometimes he fell prostrate, but was quickly up again and he finally disappeared from us, limping over the summit and the 55th saw him no more for several months."

On December 25, 1863 Howe reenlisted in the same regiment, being discharged as a corporal on November 30, 1864, and taking part in 14 battles.

Impressed with the boy's bravery and determination, the general arranged for his admission to the U.S. Naval Academy. Before this could happen,

Orion was involved in an incident near Dallas, Georgia. There, while running dispatches, he picked up a discarded rifle and fired toward a group of resting Confederates. He missed. Angered, they fired back hitting him twice in the arm and once in the chest. Finally, in 1865, Orion became a Midshipman. But it didn't last. He did not do well with the discipline and was dismissed after two years.

Orion's military adventures continued after the war. He was shipwrecked in the Merchant Marine, and wounded and left for dead in the Indian campaigns in northern California. Decades after the war, units were given the opportunity to name one of their own for the Medal of Honor. The 55th named Orion. He received his Medal of Honor on April 23, 1896.

In 1982 the Waukegan, IL National Guard Armory was renamed in his honor. The 933rd Military Police Company currently drills there.

Primary source:

G. Clifton Wisler, When Johnny Went Marching, 2001, pp. 31-34

General William Tecumseh. SHERMAN

The following is from the book called The Civil War in Song and Story, By Frank Moore, published in 1889, by P.F. Collier, Publisher. Pg. 104-

A Brave Drummer Boy- Orion P. Howe, of Waukegan, Illinios, drummer-boy to the Fifty-fifth Volunteers of that State, was appointed to fill a vacancy in the Naval School at Newport. The following extract from a letter, written by Major-General Sherman to Secretary Stanton detailing an incident which transpired during the assault upon the rebel works at Vicksburg, on May 19th, doubtless secured the boy's promotion:

"When the assault at Vicksburg was at it's height on the 19th of May, and I was in front near the road which formed my line of attack, this young lad came up to me wounded and bleeding, with a good, healthy boy's cry: 'Gen. Sherman, send some cartridges to Col. Malmborg; the men are nearly all out.' What is the matter, my boy? 'They shot me in the leg, sir, but I can go to the hospital. Send the cartridges right away.' Even where we stood, the shot fell thick, and I told him to go to the rear at once, I would attend to the cartridges, and off he limped. Just before he disappeared on the hill, he turned and called as loud as he could: 'Calibre 54.' I have not seen the lad since, and his Colonel, Malmborg, on inquiry, gives me his address as above, and says he is a bright, intelligent boy, with a fair preliminary education."

What arrested my attention then was, and what renews my memory of the fact now is, that one so young, carrying a musket-ball wound through his leg, should have found his way to me on that fatal spot, and delivered his message, not forgetting the very important part even of the calibre of his musket, 54, which you know is an unusual one.

"I'll warrant that the boy has in him the elements of a man, and I commend him to the Government as one worthy the fostering care of some one of it's national institutions."

Orion P. Howe, with his regimental commander Col. Oscar Malmborg,

John Cook, Bugler

Rank and organization: Bugler, Battery B, 4th U.S. Artillery. Place and date: At Antietam Md., 17 September 1862. Entered service at: Cincinnati, Ohio. Birth: Hamilton County, Ohio. Date of issue: 30 June 1894. Citation: Volunteered at the age of 15 years to act as a cannoneer, and as such volunteer served a gun under a terrific fire of the enemy.

George D. Sidman
the heroics of a Civil War drummer boy

In June of 1862 George D. Sidman was a 16 year old drummer boy with the 16th Michigan Infantry, Company C. In the midst of an assault at Gaines Mills, Virginia, Sidman volunteered to carry the regimental flag and rallied his comrades in the face of grave danger until he was wounded in the hip. For his distinguished bravery, Sidman was awarded the Medal of Honor.

In recounting Sidman's heroics on that day, Captain Ziba Graham stated:

"Well do I remember that December day in 1862, as we stood en masse on Stafford Heights, overlooking Fredericksburg, all ready to cross the Rappahannock, when the first brigade colors for our brigade were brought upon the field. I can see now the eagerness with which this comrade Sidman, a mere boy, with scarce the down of young manhood upon his chin, sprang forward from the ranks and begged of me the permission to carry those colors. It was granted. Colonel Stockton in command, admiring his pluck but deprecating his youth, finally gave his consent. Sidman brought them out of that hell of fire, many holes shot in them, himself wounded. On his breast to-day he wears the medal of honor, a patent of nobility for bravery far above riches and above price."

George Hollat

George Hollat, 16, was a first class boy or powder monkey on *USS Varuna* during an attack on Forts Jackson and St. Phillip in April 1862. He was responsible for running bags of powder from the store room below deck to the gun crew. His citation reads, "He rendered gallant service through the perilous action and remained steadfast and courageous at his battle station despite extremely heavy fire and the ramming of the *Varuna* by the Rebel ship, *Morgan*, continuing his efforts until his ship, repeatedly holed and fatally damaged, was beached and sunk."

Nathaniel McLean Gwynne, age 15

At age eleven, Nathaniel (Nat) Gwynne saw the world around him dissolved into war. Like so many others, he longed to be a part of it, to march off to war with the rest of the soldiers. But he was too young. Finally, late in the war, he could stand it no longer and ran off to war without his parents' permission. Units he attempted to join refused him. His state of Ohio organized its last cavalry unit, the Ohio 13th and he was allowed to sign up as a mascot of sorts without official enlistment, on May 3, 1864. A short time later, he was issued a bugle, became the bugler for Company H, and was issued a musician's uniform. Nearly three months passed and he found himself with his unit, poised for an attack outside of Petersburg, Virginia.

At 4:40am, the morning of July 30, 1864, a huge explosion opened a gaping crater in the Confederate lines as dirt, smoke, flame, and human bodies rose nearly 200 feet into the air. As the 13th prepared to advance, Nat's commanding officer reminded him that he was not an official soldier and ordered him to stay behind. The boy replied that he came to fight and fight he would, then, a half hour later, advanced toward the crater with the rest of his command, on foot in columns of fours. At the halfway point they were ordered to lie down. Their position was under a heavy fire of artillery and musketry. After ten minutes, the regiment rose up and moved forward toward the edge of the crater.

In the midst of the struggle, the regimental colors went down and the boy rushed forward to save them. On the way back, his left arm was struck by an enemy shell. Holding the flagstaff firmly with his good arm, he continued on. A short time later, he was hit through a knee. He finally reached the safety of the Federal line where, within two hours, he found himself on a surgeon's table.

Gwynne's left arm was amputated just below the shoulder. Upon examination, his knee wound was cleaned and dressed. Four days later he was sent to Douglas General Hospital in Washington, DC. The boy was transferred to Second Corps Hospital in Alexandria before month's end. His brother came in from Cincinnati to help care for him.

Sometime later in early December, someone at the war department discovered that Nat had never been officially enlisted; officially, he was a private citizen. On December 19, 1864, the oversight was corrected and his enlistment was backdated to May 10, 1864. And he received full pay for his entire time in the army.

By the end of the month, as Nat's recovery remained steady, his brother arranged for him to be transferred to a hospital in Cincinnati. Once back home, his story became known to the public resulting in a nomination for the nation's highest honor, the Congressional Medal of Honor. The application was submitted January 3, 1865.

Just weeks later, on January 27th, Nathaniel McLean Gwynne was issued the Congressional Medal of Honor with the following citation: "The president of the United States of America, in the name of Congress, takes pleasure in presenting the Medal of Honor to Private Nathaniel McLean Gwynne, United States Army, for extraordinary heroism on 30 July 1864, while serving with Company H, 13th Ohio Cavalry, in action at Petersburg, Virginia. When about entering upon the charge, this soldier, then but 15 years old, was cautioned not to go in as he had not been mustered. He indignantly protested and participated in the charge, his left arm being crushed by a shell and amputated soon afterward."

Nat left the army a month later with a disability discharge on March 21, 1865.

Source material:-

Hoar, Jay S., *Callow, Brave, and True, a Gospel of Civil War Youth*, Gettysburg, PA: Thomas Publications, 1999, pp. 22-23
Keesee, Dennis M., *Too Young to Die, Boy Soldiers of the Union Army 1861-1865*, Huntington, West Virginia, Blue Acorn Press 2001, pp.4-5
https://valor.militarytimes.com/hero/2436

WILLIAM MAGEE, AGE 15

A brave drummer boy earns the Congressional Medal of Honor

In his ***New Jersey and the Rebellion,*** John Y. Foster describes the heroism of a young drummer boy name William Magee. For his valiant efforts, Magee was awarded the Congressional Medal of Honor. Magee's citation reads, "In a charge, was among the first to reach a battery of the enemy and, with one or two others, mounted the artillery horses and took two guns into the Union lines."

Foster describes the details of drummer boy Magee's bravery in his account. He writes:

Among the many instances of youthful intrepidity and daring, none, perhaps, exceeded in all the points of real sublimity as those which are furnished in the career of drummer William Magee, of the Thirty-third Regiment. This lad, for he was only a lad, entered the service at fifteen years of age-leaving a widowed mother in the city of Newark-to aid in maintaining the unity of the Nation. From the first he displayed qualities of the highest order. Intelligent, fearless, vigilant, he was at all times an example alike to superiors and inferiors. Though entering the service as a drummer, he by no means confined himself to the duties of his specific sphere. He had a knack of fighting as well as drumming, and withal exhibited an appreciation of the methods of warfare which qualified him for the most surprising exploits. One of these, at least, was equal in splendor of execution and grandeur of result to any which the history of the war records. It will be remembered that in the fall of 1864, after Sherman had swung loose from his base and started on his stately' "March to the Sea," Hood with an army of forty thousand men laid siege to Nashville, defended by General Thomas. Here, for a period of two or three weeks, our troops were penned up with little prospect of relief.

At Murfreesboro, thirty miles away, General Thomas, reluctant to relax his hold on the railroad to Chattanooga, had stationed a small garrison under General Milroy. This garrison, as the rebels gathered in greater force, beleaguering the post, soon became comparatively isolated, all avenues of escape being practically closed. But the men did not lose heart. At length, on the 2nd of December, it was determined to strike a blow for deliverance. At this time, young Magee had become acting orderly to General VanCleve, and to him, youth as he was, the order was given to charge the enemy. It may be that a smile accompanied the order – a smile at the thought of committing such a work to a mere stripling; but it is certain that the confidence of the' commander was not misplaced. Taking the One Hundred and Eighty-first Ohio Infantry, Magee sallied out of the works, and rushed upon a battery posted on an eminence hard by. The charge was made most gallantly, but the fire of the enemy was resistless, and slowly the column fell back. But the intrepid orderly did not for a moment falter in his purpose. One repulse only stimulated his appetite for his work, and accordingly, selecting the One Hundred and Seventy-fourth Ohio, he again moved out, again charged the foe, again met their withering fire; still, however, pressing on until at last the victory was his. And it was no ordinary victory. Two heavy guns and eight hundred of the enemy killed, wounded and captured, were the trophies which he brought out of the contest. Nor was this all. This signal success at once dispiriting the enemy and reviving the hopes of our own men, proved the first of a series of victories which resulted, finally, in driving Hood from Tennessee and restoring that whole section to Federal control. The readiness and gallantry displayed by young Magee in this affair very naturally attracted the attention of those around him, and he received the hearty commendation of Generals Rosseau, Milroy, and other officers in command. Subsequently he received a medal of honor from the War Department, inscribed, "The Congress to drummer William Magee, Company C, Thirty-third Regiment, New Jersey Volunteers."

Source:

New Jersey and the Rebellion by John Y. Foster (1868)
https://civilwartalk.com/threads/willie-magee-16-year-old-medal-of-honor-hero.142364/
[NOTE:- Willie's story stood for several years, but was eventually proven to be a lie created by Willie himself and exposed by the men who were there and knew the truth.}
Drummer Boy Willie Magee, Civil War Hero and Fraud, by Thomas Fox (2007)

Remember the Boys Our School Books Ignore

WHO GAVE THE LAST FULL MEASURE DURING THE CIVIL WAR

six of the hundreds who died, of the more than 200,000 who served

Edwin Francis Jemison, 15
2nd Louisiana Infantry, killed
at the Battle of Malvern Hill

Rashio Crane, 15
Co D, 17th Wisconsin, died
in Andersonville Prison

Benjamin Knox, 15
Co. H, 20th Ohio, killed in
the trenches at Atlanta

Charley King, 12
Drum Major, Co F 49th PA
Killed at Antietam

William Hugh McDowell, 17
VMI Cadet 1st of 10
killed at New Market

Clarence McKenzie, 12
Co D, 13th NY State Malitia
accidentally killed by friendly fire

EDWIN FRANCIS JEMISON
a Confederate Drummer Boy

Private Edwin Jemison of the Georgia Infantry was killed at the Battle of Malvern Hill, Virginia, on July 1, 1862. During this battle the Union army did not use trenches. Instead they stood in battle formation. They were backed by 100 cannons in front and another 100 plus on the flank. The Confederates charged and suffered more than 5,300 casualties including Pvt. Jemison.

Pvt. Edwin Francis Jemison of the 2nd Louisiana Regiment, C.S.A., was shot and killed at Malvern Hill, Va., on July 1, 1862, at the age of 17. In July 1864, as the deaths of young boys mounted, the War Department required that officers cease enlisting soldiers under 16 years old or face severe penalties. Despite these orders, new underage recruits, many who would never return home, joined until the war's end.

http://en.wikipedia.org/wiki/Edwin_Francis_Jemison

RASHIO CRANE

Rashio Crane was a 15-year-old drummer with Company D 7th Wisconsin. He was captured May 5, 1864 at the Wilderness while helping a wounded comrade. Sent to Andersonville Prison, he took sick and died July 23, 1864.

BENJAMIN KNOX

Benjamin Knox was a 15-year-old private in Company H 20th Ohio, from Vicksburg to the Atlanta Campaign. He was shot in the trenches at Atlanta and died a short time later in the company quarters.

Charles "Charley" King

12 YEARS 5 MO. AND 9 DAYS. OLD WHEN HE ENLISTED.

Story in section about the youngest.

William Hugh McDowell

Most young readers of Civil War historic fiction meet Hugh for the first time in Elaine Marie Alphin's book ***Ghost Cadet***. In the story, Benjy, a teenager visiting his grandmother in New Market, Virginia, meets a strange boy as he visits the battlefield. It turns out that the boy is a ghost from the battle that took place there over a hundred years before and he is looking for his lost watch. It turns out through research that Hugh, his watch, and the battle are all real.

William Hugh McDowell was born December 31, 1846, at Beattie's Ford in Iredale County, North Carolina. He entered the Freshman class at Virginia Military Institute August 22, 1863, at the age of sixteen. The following letter of reference requests a cadetship for Hugh.

The following letter was written by Mary Anna Jackson, the wife of General Stonewall Jackson, on behalf of her relative William H. McDowell. The letter was addressed to General Francis H. Smith, the Superintendent (President) of VMI.

Charlotte, N.C. Jany 20th 1863
Gen'l F.H. Smith
My dear Sir,

I have been requested by a cousin of mine, Mrs. McDowell, to make application to you for a cadetship in the Institute for her son--a youth of fine character & talents. I would be much gratified if you could receive him, as I feel special interest in him, feel assured he will do well & his parents are very anxious for him to have a military education. Please let me hear from you as early as you conveniently can on the subject.

My love to Mrs. Smith & your daughters. With kind regards for yourself, I am, my dear Gen'l,

Yours very truly, Mary Anna Jackson.

Many cadets and instructors from VMI left as the Civil War got underway to serve in the Confederate Army. One was Mary Anna Jackson's husband, General "Stonewall" Jackson. In the following letter, Hugh's mother writes to request information about arrival time at the institute and to introduce her son to the superintendent.

The following letter was addressed to Francis H. Smith, the Superintendent (President) of VMI. The cousin she refers to is Mary Anna Jackson, the widow of General Stonewall Jackson.

June 1st, 1863
Col. Smith

I write for information concerning the time when the exercises at Lexington commence; as you promised Mrs. Jackson last winter that you would take him in July but did not state what time in July. Please let me know the time, the regulations, and terms.

I might obtain the necessary information from my cousin but her grief is too recent, too great, & sacred to obtrude upon with my concerns. Virginia had reason to be proud of & thankful for such a chieftain as Jackson. A great & good man, a pure & unselfish patriot, and it is a pleasure to us to think that we can do something by kind offices & soothing attentions, to cheer his widow on her lonely way. We all mourn him, through the length & breadth of this Confederacy.

You will find our child careless & thoughtless, but high principled, & too firm to be led astray. I hope his conduct & deportment may be unexceptionable as it has been hitherto. And let me beg of you to take an interest in him. I can scarce hope to have him with me much more after he goes to you--as when he leaves you twill be to enter the army. He has been a good obedient child to me and I would feel relieved to know that far from home and among strangers he has found one friend and protector.

Please direct your letter to Mrs. R.R. McDowell, Mt. Mourne, P.O. Iredell Cty, N.C.

Very respectfully, R.R. McDowell

In the fall of the year, Mrs. McDowell sent a letter with money to request that her son's picture be taken.

The following letter was addressed to Francis H. Smith, the Superintendent (President) of VMI.

Oct 3rd 1863
Genrl Smith

I enclose you $10 for the use of my son William H. McDowell, with which I beg that you will have a good Daguerreotype, or photograph of him taken. He is my eldest child, and is far from me. And should any misfortune befall him, I would wish some likeness of him preserved. I have no idea of the cost of such a thing at the present time, and should this be insufficient for the purpose I will remit more. Should it be more than enough, tis subject to your discretion. Willie can retain the daguerreotype until there is an opportunity of sending it to us. By attending to this request, you will confer a favor on

R.R. McDowell

Throughout the war, VMI's superintendent, General Francis S. Smith, had made known the availability of the institute's corps of cadets with supporting artillery, in the event they were needed to help defend the Shenandoah Valley region. He was thanked and kindly told that it would not be necessary. In May of 1863, General Stonewall Jackson, an instructor from the institute, was mortally wounded at the Battle of Chancellorsville. A year later, May 10, 1864, cadets from VMI participated in a graveside memorial.

That night the VMI cadet corps was called out to report to General Breckenridge at Staunton to help defend against a Union army invading from the north. William Hugh McDowell along with 256 fellow cadets and a battery of artillery began the march which would lead to New Market. As they marched north from Lexington, they spent the nights in the open or in a church along the way, constantly exposed to the weather which included much rain. Arriving at New Market on the eve of battle and in a rainy downpour, the VMI cadet corps was placed in reserve behind the front lines. They were awakened around one AM in a pouring rain to advance to their position. In their approach to the battle, they had to descend a hill which they did in parade field formation, and came under immediate artillery fire from the enemy. Three cadets were killed here, Cabell, Crockett, and Jones, all killed by the same shell.

"A little removed from the spot where Cabell fell, and nearer to the position of the enemy, lay McDowell, it was a sight to wring one's heart. That little boy was lying there asleep, more fit, indeed, for the cradle than the grave. He was barely sixteen, I judge, and by no means robust for his age. He was a North Carolinian. He had torn open his jacket and shirt, and, even in death, lay clutching them back, exposing a fair breast with its red wound." (From *An End of an Era* by John S. Wise, copyright 1899)

Ten days later this letter was sent to Hugh's father.

May 25, 1864. Death of Cadet William H. McDowell
Virginia Mil. Institute
May 25th, 1864
Mr. R.I. McDowell
Mount Mourne, Iredell Co. N.C.

Sir--

You have doubtless received before this the mournful intelligence that your [noble] son has been added to the long list of the gallant dead who have fallen in defending their country against the invasion of a ruthless foe. The newspapers have furnished you with accounts of the victory gained by Gen. Breckinridge over Sigel near New Market, and every notice of the fight bears unequivocal testimony to the value of the aid rendered by the Corps of Cadets and the [illegible] valour that they displayed in the action. You have also received, I suppose, an official letter from the Adjutant informing you of the sad event.

I can add nothing more except the statement that the fatal ball passed entirely through his body, entering a little to the [illegible] of the breastbone and coming out on the left of the spine, passing probably through the heart, so that it may be concluded that his death was instantaneous.

This I received from Col. Gilham who examined the body before its interment. I have not been able to see anyone who was near him when he fell, as the cadets have not returned to the Institute, [having been] ordered to Richmond.

The Quartermaster will endeavor to preserve any mementos or any property of the cadets who have fallen, but cannot at present while the Corps is absent identify what belongs to each. The letter which you gave me for him [3 words illegible] and which weighed as a heavy burden on my heart after I heard before I reached home, that the words of affection it contained could never reach the eyes closed in death forever--together with a second one received from the office for him, I have directed to be kept subject to your order not choosing to subject them to the risk of the mail in the present uncertainty of transmission.

I offer no words of condolence. I know how to sympathize with you for my noblest son fell slain in battle not two months after he left the Institute--and I know by experience that the only comfort for so great a sorrow must come from a source higher than any on earth.

Yrs truly
J.T.L. Preston

Letter of reply from Robert McDowell regarding the death of William.

Mount Mourne, May 30 1864 N.C.
J.N. Morrison, Esq.
Sir:

   Your letter informing me of the death of my son Wm McDowell has been received. It came upon me like a clap of thunder in a clear sky, as I was not aware the cadets had been called out. I wrote a letter to Col. Preston in regard to his clothes, books, and gold watch, desiring him to have them sent to me at Charlotte by Express. I would feel greatly obliged to you for any assistance rendered in this matter. I desire to retain them as memorial of my beloved son, [thus] cut off in the opening of life.

Respectfully
R.I. McDowell

In all, five cadets were killed in battle and five more died later. Breckinridge was victorious and the cadets played a major role in the action, charging across a field of mud and capturing enemy artillery.

Hugh and the other four who died that day were buried in the churchyard in New Market. They were later reburied in a small cemetery at VMI. Years later, a memorial was placed at the Virginia Military Institute for the cadets who died in battle at New Market. The bodies of the six cadets whose markers can be seen behind the memorial are buried beneath the monument in a copper box. The memorial was created by Cadet Moses Ezekiel, the first Jewish cadet at VMI, who fought that day at New Market and who sat with his mortally wounded friend Cadet Jefferson until he died and who wanted to be an artist when he grew up.

Finally, this letter from Hugh's mother describes her grief following his death.

This letter is courtesy Dr. and Mrs. Murphy Cronland, owners of the original document, and reproduced with their permission

July 25th, 1864
My Dear Aunt,

I have been intending to write to you but have felt so badly that I put it off from day to day, hoping that my heavy sorrow would grow lighter. But it does seem to me that it only deepens as I reflect on it and realize my loss as days pass away.

At first, I could think of every blessing that had been vouchsafed me in connection with it. But now, altho' I do not murmur or complain and can from the heart say "Thy will be done," yet I recall the mercies remaining, but can't feel the same gratitude.

I felt so thankful that the poor, dear child was not wounded & taken prisoner by our cruel foes--that he did not linger in agony and that as he had to die, that he died in the discharge of his duty to his God & his country, and not a craven coward, [illegible] away his life. I realized that there are some things, harder to bear than death, the disgrace of those we love. I felt so thankful that I had full assurance of happiness & that he passed from earth to Heaven with but one sigh.

But now I only feel my loss. I can't think of him in Heaven with that bright angelic host mingling his praises with the Redeemed. I only feel that his loss to me is [irreparable]. That I shall no more see his form -- so erect --no more gaze at his beautiful eyes--lit up with mirth or enthusiasm -- no more see those dimples in his cheeks as he would break out into his merry peal of laughter or look at the long dark lashes when he was in thought. This was the month in which he was to have been at home, and when I had expected to send him into the army with my prayers & blessings. I had made up my mind to give him to his God & country, but O not so soon! Could I weep, it would bring relief, but I cannot.

Before these Union people, some of whom I have heard of exulting him his death, I talk of my noble hero boy. I am calm - cheerful. I tell how thankful I am that he fell at the post of honor & duty & c. But my heart - O how it aches! afterwards - but if I died, they should not know if was with grief. My child died in defence of the South. To that cause my life is devoted and my God in his mercy take all that are dear to me & myself before we ever bend to yankee rule.

We had a kind & sympathizing letter from a gentleman in Lex[ington] whose son was a room mate of Willie's on last Saturday. He says that his son Edward Tutwiler & Willie were "fast friends" & that his son was much attached to him - that a short time before the Yankees took possession of Lexington he visited the room the boys occupied for the purpose of getting away the clothes & other things left by his son & on examination he found several articles belonging to our son, among them his daguerreotype. He says that he looked for his trunk but it could not be found & Col Preston writes that it is believed to have been burnt in the Inst or carried off by free negroes before.

His watch (his father's gold one, bought while he was in college) cannot be heard of, neither his Bible. Col Preston's son, who was Capt of the Company, said that he assisted in burying him, that there was no mutilation, no bruise on the body except where the fatal ball entered but neither his watch nor Bible were on his person. Mr. Tutwiler says that his son in writing to him from New Market said "my roommate McDowell was killed, in the front rank. I know he has gone to heaven for he was a sincere Christian" & the Father adds in his letter --"This should cheer you, in your sad affliction, for your loss has been His gain."

Revd Dr. White of Lexington, whom I've met with at the Gen'l [illegible], has sent us part of his hair, retaining the other half for fear it might be lost, and writes that he hears Willie spoken of by every body who knew him in the most flattering terms--by Profs. & c. Dr. White & Col. Preston have both lost most promising sons in battle, & seem to sympathise much with us. We have been treated with much kindness by all our friends & had very many kind letters.

Willie had not been social until grace sanctified his heart. But on his last visit home his health was better & he was so bright & merry and his associates always spoke of him as being so intelligent, & well read, so truthful & reliable, despising everything low & mean. He had the qualities that would have ensured success--integrity, perseverance & energy with great [several words illegible]....

...We had such a kind letter from dear James about Willie. May God spare him to you, dear Aunt & bless you all. With much love to Lizzie & Carrie, yr attached niece, R A McD

Much more can be learned about William Hugh McDowell and the corps of cadets who fought in the battle at New Market as well as the battle itself from the following websites:

http://www.vmi.edu/archives.aspx?id=5187 VMI archives
http://www2.vmi.edu/museum/nm/   New Market Battlefield State Historical Park

CLARENCE MCKENZIE, DRUMMER BOY WAS 1ST BROOKLYN CIVIL WAR DEATH
12-Year-Old Inspired Other Soldiers

A sketch of Clarence McKenzie, a drummer boy who was the first Civil War casualty from Brooklyn. He died by accident, but the 12-year old boy was remembered with reverence by his fellow soldiers, family, friends, classmates, his dog and history.

Clarence McKenzie, the drummer boy of Company D, 13th New York State Militia regiment, died after suffering a gunshot wound on June 11, 1861. He was the first Brooklyn casualty of the Civil War and the youngest fatality, though not in combat.

"He was the smallest in the corps and liked by everyone who knew him, being well behaved, always in good spirits, and ready and willing to do whatever was asked of him," an officer in the regiment recalled. "His comrade drummers and drum major were very much affected; they could not have felt worse had he been their own brother." Rev. Luther Goodyear Bingham compiled this recollection and others into his book about McKenzie, entitled The Little Drummer Boy.

McKenzie joined the regiment on July 9, 1860, and played at the Prince of Wales' reception in October. The Civil War began with the Confederate attack on Fort Sumter on Apr. 12, 1861, and McKenzie's regiment was shipped to a Union military facility in Annapolis, Maryland on Apr. 30.

McKenzie's teachers gathered and bid him a tearful farewell, and his fellow soldiers aboard the ship were struck dumb by the sight of McKenzie kneeling to pray before bed.

"I was deeply affected, and the thought came rushing into my mind, 'that is probably the way in which my praying wife is praying for me, at this very moment,'" a soldier recalled. "Many a heart was compelled to feel, and many an eye glistened with tears."

McKenzie's brother William was also a drummer boy in the regiment, and their letters home were full of good cheer, admonitions not to worry and requests.

"Please, dear mother, send me on another cake," McKenzie said in one post script. "The boys took it all from me that is a dear good mother. Your Clarry." In another post script, McKenzie promised to send his mother a cracker from his rations, "for a sample of what we get."

Though homesick, McKenzie wrote that he was enjoying the weather, his surroundings and the company of his fellow soldiers.

"Dear mother, do not cry for me, for I am well off, and I hope to return to you in three months or sooner," McKenzie wrote in his last letter, dated May 28.

McKenzie was sitting near a wall in the drummer's quarters when the other soldier mistakenly shot him with a musket he'd borrowed. The soldier didn't know the weapon was loaded, according to the regiment's investigation, and he wasn't charged with any wrongdoing.

"A few moments previous to drill he was practicing the manual in the drummer's quarters, and in coming to a charge bayonet his hand struck the hammer of his piece, forcing it down although he says it was half cocked and discharging it, the ball striking Clarence McKenzie in the back, passing through and out at the stomach, and finally striking against a brick wall with such force as to break out part of the brick," an officer in the 13th Regiment reported.

Soldiers gathered around B Company Quarters where McKenzie lay inside, dying. He expressed forgiveness for the soldier who shot him, and when he calmly passed away about two hours later, the soldiers "wended their way to their rooms with saddened hearts," the officer reported.

"The drum major was almost heartbroken," another soldier re called. "The gloom cast over the camp was plainly visible in every face."

McKenzie's body was packed in ice, given a military escort back to Brooklyn and buried there with full military honors. Mourners packed the church and the streets outside, and McKenzie's terrier laid down on the earth after it was mounded over his grave.

"For many nights afterwards he was in the habit of going and spending part of the night upon the grave," Bingham said of the dog, "and toward morning he would return to the house where he belonged."

More than 500,000 New Yorkers enlisted in the Army and Navy during the four years of the Civil War and 53,114 New Yorkers died.

For more information go to www.dmna.state.ny.us/civilwar

CHARLIE COULSON, A DRUMMER BOY:
A TRUE STORY IN THE AMERICAN CIVIL WAR
by Max Louis Rossvally, 188?

TWO or three times in my life God in His mercy touched my heart, and twice before my conversion I was under deep conviction.

During the American war, I was a surgeon in the United States army; and after the battle of Gettysburg, there were many hundred wounded soldiers in my hospital, amongst whom were twenty-eight who had been wounded so severely that they required my services at once, --some whose legs had to be amputated; some, their arms; and others, both their arm and leg. One of the latter was a boy who had been but three months in the service;

and being too young for a soldier, had enlisted as a drummer. When my assistant surgeon and one of my stewards wished to administer chloroform previous to the amputation, he turned his head aside and positively refused to receive it. When the steward told him that it was the doctor's orders, he said, "Send the doctor to me." When I came to his bedside, I said, "Young man, why do you refuse chloroform? When I found you on the battle-field, you were so far gone that I thought it hardly worth while to pick you up; but when you opened those large blue eyes, I thought you had a mother somewhere who might at that moment be thinking of her boy. I did not want you to die on the field, so ordered you to be brought here; but you have now lost so much blood that you are too weak to endure an operation without chloroform, therefore you had better let me give you some." He laid his hand on mine, and looking me in the face, said,--

"Doctor, one Sunday afternoon, in the Sabbath-school, when I was nine and a half years old, I gave my heart to Christ. I learned to trust Him then; I have been trusting Him ever since, and I can trust Him now; He is my strength and my stimulant; He will support me while you amputate my arm and leg."

"Won't you at least take some brandy?" I begged.

Again, he looked at me and said, "Doctor, when I was about five years old, my mother knelt by my side with her arms around my neck and said: 'Charlie, I am now praying to the Lord Jesus that you will never know the taste of strong drink. Your father died a drunkard, and I've asked God to use you to warn young people against the dangers of drinking.' I am now seventeen years old and I have never had anything stronger than tea or coffee. I am in all probability going to die and go into the presence of my God. Would you send me there, smelling of brandy?"

I will never forget the look he gave me. At that time I hated Jesus, but I respected that boy's loyalty to his Savior. When I saw how he loved and trusted Him to the very end, something deeply touched my heart. Despite the urgency of the moment and all the misery around, I did for that boy what I had never done for any other soldier. I asked him if he wanted to see a chaplain.

"Oh, yes, sir!" was his answer.

When the chaplain came, he recognized the young drummer from his tent prayer meetings. Taking his hand, he said, "Charlie, I'm so sorry to see you in this sad condition."

"Oh, I'm all right, sir," he answered. "The doctor offered me chloroform, but I declined it. Then he wanted to give me brandy, which I didn't want either. So now, when my Savior calls me, I can go to Him in my right mind."

"You might not die, Charlie," said the chaplain, "but if the Lord should call you home, is there anything I can do for you after you're gone?"

"Chaplain, please put your hand under my pillow and take my little Bible. In it you will find my mother's address; please send it to her, and write a letter, and tell her that since the day I left home I have never let a day pass without reading a portion of God's Word, and daily praying that God would bless my dear mother,-- no matter whether on the march, on the battlefield or in the hospital."

"Is there any thing else that I can do for you, my lad?" asked the chaplain.

"Yes; please write a letter to the superintendent of the Sands Street Sunday School, Brooklyn, N.Y.; and tell him that the kind words, many prayers, and good advice he gave me I have never forgotten; they have followed me through all the dangers of battle, and now, in my dying hour, I ask my Savior to bless and strengthen my dear old teacher: that is all."

Turning toward me, he said, "Now doctor, I am ready; and I promise you that I will not even groan while you take off my arm and leg, if you will not offer me chloroform."

I promised, but I had not the courage to take the knife in my hand to perform the operation without first going into the next room and taking a little stimulant to nerve myself to perform my duty.

While cutting through the flesh, Charlie Coulson never groaned, but when I took the saw to separate the bone, the lad took the corner of his pillow in his mouth, and all that I could hear him utter was, "Oh, Jesus, blessed Jesus, stand by me now." He kept his promise, and never groaned.

I couldn't sleep that night. Despite the constant moans and weeping of the wounded, all I could see was Charlie's soft blue eyes. Even his words, "Blessed Jesus, stand by me now," kept ringing in my ears. Between twelve and one o'clock, a strong urge came over me to see that boy again. When I got there, I was told that sixteen of the badly wounded had died. "Was Charlie Colson one of them?" I asked. "No, sir," answered the steward, "he's sleeping as sweetly as a babe."

When I came to his side, one of the civilian nurses informed me that at about nine o'clock, two members of the U.S. Christian Commission, accompanied by the chaplain, came to read Scripture and sing hymns. She said that the chaplain knelt by Charlie and offered up a passionate prayer. Then they and Charlie sang the sweetest of all hymns, "Jesus, Lover Of My Soul." I couldn't understand how this young lad, who suffered such horrible pain, could sing.

Five days after his amputation, Charlie sent for me. "Doctor," he said, "my time has come. I don't expect to see another sunrise. But thank God, I have no fear and I'm ready to go. I want to thank you with all my heart for your kindness to me. I know you are Jewish, and that you don't believe in Jesus, but I want you to stay with me, and see me die trusting my Savior to the last moment of my life."

I tried to stay, but I could not. I didn't have the courage to stand by and watch a true Christian die, rejoicing in the love of Jesus whom I had been taught to deny. So, I hurriedly left.

About twenty minutes later an anxious steward found me and said, "Doctor, Drummer Colson wants to see you again." "I've just seen him," I answered, "I can't see him again." "But, Doctor, he says he must see you before he dies."

So, I made up my mind to see him, say a few kind words, and let him die. However, I was determined that no talk about his Jesus was going to influence me.

His condition had worsened. Asking me to take his hand, he said, "Doctor, I love you because you are Jewish; the best friend I have found in this world was also Jewish."

I asked him who that was, and he answered, "Jesus the Christ, and I want to introduce you to Him before I die. Will you promise me, Doctor, that what I am about to say to you, you will never forget?"

I promised, and he said, "Five days ago, while you operated on me, I prayed to the Lord to save you."

His words pierced deep into my heart. I couldn't understand how, when I was causing him the most intense pain, he could forget all about himself and think only of his Savior and my spiritual need. All I could say at the moment was, "Well, my dear boy, you will soon be all right."

I started to leave, hearing him sing softly, "I'm going home to die no more."

Twelve minutes later he fell asleep, "safe in the arms of Jesus."

Hundreds of soldiers died in my hospital during the war, but I only followed one to the grave, and that one was Charlie Coulson, the drummer-boy, and I rode three miles to see him buried. I ordered that he'd be dressed in a new uniform and like the burial for an officer, placed in an officer's coffin, and arranged that his coffin be covered with the flag he nobly served.

His dying words made a deep impression upon me. I remembered thinking how gladly I would have given all I possessed, if I could have felt towards Jesus as he did. But with the continuation of the cruel war and my company with worldly officers, I gradually forgot his prayer and my promise.

After the war and for nearly ten years, I fought against believing in Christ. But God continued to bring faithful and godly people into my life that spoke of Jesus' love. Finally, the drummer boy's prayer was answered and I accepted Jesus Christ as my personal Savior and Messiah. It did come at a high cost. My family, in-laws and dear mother rejected me. Psalm 27:10, was a great comfort, "When my father and mother forsake me, then the Lord will take me up."

It was eighteen months after my salvation that the Lord had a special blessing for me. One evening while traveling through Brooklyn, I felt led to attend a prayer meeting in a small local church. It was a meeting when believers testify to the loving-kindness of the Lord. After several had spoken, an elderly lady stood up and said,

"Dear friends, this may be the last time I have a chance to publicly share how good the Lord has been to me. My doctor told me yesterday that my right lung is nearly gone, and my left lung is failing fast, so at best, I only have a short time to be with you. But what is left of me belongs to Jesus. It's a great joy to know that I shall soon meet my boy with Jesus in heaven. My son was not only a soldier for his country, but also a soldier for Christ. He was wounded at the battle of Gettysburg, and was cared for by a Jewish doctor, who amputated his arm and leg. He died five days after the operation. The chaplain of the regiment wrote me a letter, and sent me my boy's Bible. I was told that in his dying hour, my Charlie sent for that Jewish doctor, and said to him, `Doctor, before I die I wish to tell you that five days ago, while you operated on me, I prayed to the Lord to save you."

As I heard this lady speak, I just couldn't sit still! I left my seat, ran across the room, took her hand and said, "God bless you, my dear sister. Your son's prayer has been heard and answered! I am the Jewish doctor that your Charlie prayed for, and his Savior is now my Savior!"

"Having been frequently asked whether all the details of this story are strictly true, I take this opportunity of stating that every incident occurred exactly as related." Max L. Rossvally

MICHIGAN'S LITTLE DRUMMER
BOYS OF THE CIVIL WAR
CHARLEY GARDNER

RHYTHM SECTION:
Michigan's intrepid drummer boys played
their part in beating the enemy during the Civil War
By Richard Bak, edited

Michigan troops crossing the Rappahonnock River
at Fredericksburg, Va., Dec. 11, 1862.
image courtesy of Richard Bak

Charles Gardner grew up in Flint, Michigan and was just thirteen when his father joined the Second Michigan Infantry. Soon after his favorite teacher, S.C. Guild, signed up with the Eighth Michigan Volunteer Infantry. Seeing these two men head off to serve their country made Charles feel as though he too should be helping the cause. The Flint youngster, eager to show the upstart Confederacy a thing or two, begged his mother to allow him to follow his favorite teacher and his father. Since his father, Charles, was serving with the 2nd Michigan, so his mother was reluctant to see him go as well. Charley, already a good drummer, pleaded with his mother to let him volunteer and "take the place of a man who can handle a musket." His mother eventually did consent, and Charley soon after joined the Eighth Michigan – Wandering Regiment. He enlisted as a drummer boy in Company A of the 8th Michigan, commanded by Captain Guild.

As his regiment was on its way to Port Royal, South Carolina, Charley and his father were reunited in Washington, DC. This chance encounter was the last time the two saw each other—later that same year, in Alexandria, Virginia, his father died of typhoid fever. Charley's teacher, Captain Guild, died in the battle at James Island, South Carolina, June 16, 1862. The excitement and naiveté with which the schoolboy went to war quickly faded. By the end of 1862, both of Charlie's father figures were dead. Charley stayed with the regiment, enduring long marches, short rations and weeks of being besieged in Knoxville, Tennessee. During the siege of Knoxville, Charley was wounded. The regiment was recalled to Detroit, but for the Gardner family there was no happy homecoming at the train station. Charlie, who was thought to be recovering nicely, had died en route. He was 14 years old.

Little Eddie the Drummer Boy

A story from the US Civil War

LITTLE EDDIE THE DRUMMER-BOY.
A Reminiscence Of Wilson's Ckeek.

A few days before our regiment received orders to join General Lyon, on his march to Wilson's Creek, the drummer of our company was taken sick and conveyed to the hospital, and on the evening preceding the day that we were to march, a negro was arrested within the lines of the camp, and brought before our captain, who asked him "what business he had within the lines ? " He replied: "I know a drummer that you would like to enlist in your company, and I have come to tell you of it." He was immediately requested to inform the drummer that if he would enlist for our short term of service, he would be allowed extra pay, and to do this, he must be on the ground early in the morning. The negro was then passed beyond the guard.

On the following morning there appeared before the captain's quarters during the beating of the reveille, a good looking, middle aged woman, dressed in deep mourning, leading by the hand a sharp, sprightly looking boy, apparently about twelve or thirteen years of age. Her story was soon told. She was from East Tennessee, where her husband had been killed by the rebels, and all their property destroyed. She had come to St Louis in search of her sister, but not finding her, and being destitute of money, she thought if she could procure a situation for her boy as a drummer for the short time that we had to remain in the service, she could find employment for herself, and perhaps find her sister by the time we were discharged.

During the rehearsal of her story the little fellow kept his eyes intently fixed upon the countenance of the captain, who was about to express a determination not to take so small a boy, when he spoke out: "Don't be afraid, captain, I can drum." This was spoken with so much confidence, that the captain immediately observed, with a smile : "Well, well, sergeant, bring the drum, and order our fifer to come forward." In a few moments the drum was produced, and our fifer, a tall, round shouldered, good natured fellow, from the Dubuque mines, who stood, when erect, something over six feet in height, soon made his appearance.

Upon being introduced to his new comrade, he stooped down, with his hands resting upon his knees, that were thrown forward into an acute angle, and after peering into the little fellow's face a moment, he observed :

" My little man, can you drum V "
" Yes, sir," he replied, "I drummed for Captain Hill in Tennessee."

Our fifer immediately commenced straightening himself upward until all the angles in his person had disappeared, when he placed his fife at his mouth, and played the "Flowers of Edenborough," one of the most difficult things to follow with the drum that could have been selected, and nobly did the little fellow follow him, showing himself to be a master of the drum.

When the music ceased, our captain turned to the mother and observed :

" Madam, I will take your boy. What is his name ? "

"Edward Lee," she replied; then placing her hand upon the captain's arm, she continued, "Captain, if he is not killed " — here her maternal feelings overcame her utterance, and she bent down over her boy and kissed him upon the forehead. As she arose, she observed: "Captain, you will bring him back with you, won't you ?"

"Yes, yes," he replied, "we will be certain to bring him back with us. We shall be discharged in six weeks."

In an hour after, our company led the Iowa First out of camp, our drum and fife playing "The girl I left behind me." Eddie, as we called him, soon became a great favorite with all the men in the company. When any of the boys had returned from a horticultural excursion, Eddie's share of the peaches and melons was the first apportioned out. During our heavy and fatiguing march from Rolla to Springfield, it was often amusing to see our long legged fifer wading through the mud with our little drummer mounted upon his back, and always in that position when fording streams.

During the fight at Wilson's Creek I was stationed with a part of our company on the right of Totten's battery, while the balance of our company, with a part of the Illinois regiment, was ordered down into a deep ravine upon our left, in which it was known a portion of the enemy was concealed, with whom they were soon engaged. The contest in the ravine continuing some time,

Totten suddenly wheeled his battery upon the enemy in that quarter, when they soon retreated to the high ground behind their lines. In less than twenty minutes after, Totten had driven the enemy from the ravine, the word passed from man to man throughout the army, "Lyon Is killed!" and soon after, hostilities having ceased upon both sides, the order came for our main force to fall back upon Springfield, while a part of the Iowa First and two companies of the Missouri regiment were to camp upon the ground and cover the retreat next morning.

That night I was detailed for guard duty, my turn of guard closing with the morning call. When I went out with the officer as a relief, I found that my post was upon a high eminence that overlooked the deep ravine in which our men had engaged the enemy, until Totten's battery came to their assistance.

It was a dreary, lonesome beat. The moon had gone down in the early part of the night, while the stars twinkled dimly through a hazy atmosphere, lighting up imperfectly the surrounding objects. Occasionally I would place my ear near the ground and listen for the sound of footsteps, but all was silent save the far off howling of the wolf, that seemed to scent upon the evening air the banquet that we had been preparing for him.

The hours passed slowly away, when at length the morning light began to streak along the eastern sky, making surrounding objects more plainly visible, Presently I heard a drum beat up the morning call. At first I thought it came from the camp of the enemy across the creek; but as I listened, I found that it came up from the deep ravine; for a few minutes it was silent, and then as it became more light I heard it again. I listened — the sound of the drum was familiar to me — and I knew that it was our drummer boy from Tennessee beating for help the reveille.

I was about to desert my post to go to his assistance, when I discovered the officer of the guard approaching with two men. We all listened to the sound, and were satisfied that it was Eddie's drum. I asked permission to go to his assistance. The officer hesitated, saying that the orders were to march in twenty minutes. I promised to be back in that time, and he consented.

I immediately started down the hill through the thick undergrowth, and upon reaching the valley I followed the sound of the drum, and soon found him seated upon the ground, his back leaning against the trunk of a fallen tree, while his drum hung upon a bush In front of him, reaching nearly to the ground.

As soon as he discovered me he dropped his drumsticks and exclaimed, " O Corporal ! I am so glad to see you. Give me a drink," reaching out his hand for my canteen, which was empty.

I immediately turned to bring him some water from the brook that I could hear rippling through the bushes near by, when, thinking that I was about to leave him, he commenced crying, saying: "Don't leave me. Corporal, I can't walk."

I was soon back with the water, when I discovered that both of his feet had been shot away by a cannon ball. After satisfying his thirst, he looked up into my face and said : "You don't think I will die, Corporal, do you? This man said I would not — he said the surgeon could cure my feet."

I now discovered a man lying in the grass near him. By his dress I recognized him as belonging to the enemy. It appeared that he had been shot through the bowels, and fallen near where Eddie lay. Knowing that he could not live, and seeing the condition of the boy, he had crawled to him, taken off his buckskin suspenders, and corded the little fellow's legs below the knee, and then laid down and died.

While he was telling me these particulars, I heard the tramp of cavalry coming down the ravine, and in a moment a scout of the enemy was upon us, and I was taken prisoner. I requested the officer to take Eddie up in front of him, and he did so, carrying him with great tenderness and care. When we reached the camp of the enemy the little fellow was dead.

The Indiana Democrat (Indiana, Pennsylvania) Jul 10, 1862
*****p://dmna.state.ny.us/civilwar
In the following book, you can read about "the littlest hero of the war," Eddie Lee :

Title: Brave Deeds of Union Soldiers (pg 63 – Google book LINK)
Author: Samuel Scoville
Publisher: G. W. Jacobs & company, 1915

Regimental Fife and Drum Corps (Image from www.civilwarphotogallery. com)

Civil War Sources (Link to posts tagged Drummer Boys) is a blog that uses primary documents as sources for its Civil War posts. While they don't seem to have posted anything about Eddie Lee, they have covered several other drummer boys, including Johnny Clem, who I have also posted about previously.

Henry Burke, drummer boy at Shiloh

Among the many who fell on the Federal side was little Henry Burk, the drummer boy. Some soldier who, perhaps viewed the dying scene of this brave boy, penned the following lines shortly after the battle, entitled *The Drummer Boy of Shiloh*.

THE DRUMMER BOY OF SHILOH.

On Shiloh's dark and bloody ground
the dead and wounded lay,
Amongst them was a drummer boy
that beat the drum that day;

A wounded soldier raised him up —
his drum was by his side –
He clasped his hands, and raised his eyes,
and prayed before he died.

"Look down upon the battlefield,
O Thou our heavenly Friend,
Have mercy on our sinful souls" —
the soldiers cried, "Amen!"

For gathered 'round, a little group,
each brave man knelt and cried –
They listened to the drummer boy
who prayed before he died.

"Oh, Mother," said the dying boy,
"Look down from Heaven on me!
Receive me to thy fond embrace!
Oh, take me home to thee!

"I've loved my country as my God,
to serve them both I've tried,"
He smiled, shook hands, death seized the boy
who prayed before he died.

Each soldier wept then like a child —
stout hearts were they, and brave;
The flag, it was his winding sheet –
they laid him in his grave.

One wrote upon a simple board
these words, "This is a guide,
To those who mourn the drummer boy
who prayed before he died."

Angels round the throne of grace,
look down upon the brave,
Who fought and died on Shiloh's plain
now slumb'ring in the grave.

DAVID O. DODD,
BOY MARTYR OF ARKANSAS, AGE 17

On January 8, 1864, a seventeen year old boy was executed by Federal forces on the grounds of St. John's Masonic School in Little Rock.

This boy, David O. Dodd, was born on November 10, 1846 in Victoria, Texas. The only son of Andrew Marion Dodd and Lydia Echols Dodd. He had two sisters, Senhora and Lenora. In the spring of 1858 the family moved to Benton, Arkansas and then in 1861 or early 1862 to Little Rock. David enrolled in the St. John's Masonic College at that time. He attended the school for a short while and was quite popular until forced to take a sick leave for malaria.

After leaving school David obtained employment with a telegraph office in Little Rock until summer of 1862 when he accompanied his father to Monroe, Louisiana where he obtained similar employment. He remained in Monroe for a short time though he missed his home and sisters dearly.

In November of 1862 David accompanied his father to Jackson, Mississippi where he had been hired as a sutler for the Third Arkansas Dismounted Rifles near Granada. Here David stayed with a black servant, Ben, while his father followed the army. His time in Mississippi was lonely as evidenced by his letters home.

On his return his father sent his son back to Arkansas to visit his relatives and friends and to take care of some family business. His father entrusted him to sell tobacco and other merchandise to the Confederate soldiers among whom David developed friendships.

The fall of 1863 found the Federal Army in control of the city of Little Rock and David's father asked him to bring his mother and sisters to Jackson, Mississippi. David convinced his family to leave but some disturbance on the boat between Mrs. Dodd and Federal soldiers prevented their journey.

David took up employment with a sutler for the Federal troops and remained so employed until his father arrived and spirited his family away to the city of Camden, still within friendly lines.

At some point after their arrival in Camden, the elder Dodd sent his son back to Little Rock to tie up loose ends. On the journey back to the capital city he went to the Confederate headquarters at Princeton to secure a travel pass. General Fagan issued the pass on December 22.

David arrived in Little Rock, visited with some female acquaintances and delivered some letters for his sisters to their friends, and finished up his father's business seeking investors in a tobacco purchase.

On the morning of December 29, David started out for Camden on a mule with a Federal pass that he had obtained from the Federal provost marshal. When he reached the city limits his pass was confiscated and he was sent on his way south. David then made a detour to visit his uncle in Benton. The next morning he set out again and wound up back inside Union held territory and confronted by another picket. Since he had no longer had a pass he was arrested and sent to regimental headquarters. When he was asked for some form of identification he handed over his small memo book. The officer noticed some marks in the back of the book that looked like morse code. He translated the first line which said "3rd Ohio Battery has 4 guns, brass."

David was charged with spying and held until a military tribunal was formed to try him. Dodd was pronounced guilty and sentenced to hang on January 8 at St. John's Masonic College.

Many friends of the Dodds went to General Steele and begged for David's life to no avail. On January 8th Dodd wrote this letter to his family:

My Dear Parents and Sisters

I was arrested as a Spy and tried and was sentenced to be hung today at 3 o'clock the time is fast approaching but thank God I am prepared to die. I expect to meet you all in heaven do not weep for me for I will be better off in heaven. I will soon be out of this world of sorrow and trouble. I would like to see you all before I die but let God's will be done not ours. I pray to God to give you strength to bear your troubles while in this world. I hope God will receive you in heaven. Mother I know it will be hard for you to give up your only son but you must remember it is God's will. Good by. God will give you strength to bear your troubles. I pray that we may meet in heaven. Good by. God will bless you all. Your son and brother.

David O. Dodd

Early in the afternoon on January 8, a crowd of about 6,000 gathered on the grounds of the college and watched as David O. Dodd sat on his coffin on the back of a wagon. The wagon was backed up under the noose and David was made to stand on the tailboard. The noose was placed around his neck and a blindfold made of his own handkerchief. An invocation was given and the cord holding up the tailboard was cut. The rope was, apparently, incorrectly placed causing David to strangle to death.

A Union soldier who witnessed this sad spectacle wrote:

It was my unpleasant duty during the winter of 1864 to be present as one of the guards at the military execution of young David O. Dodd as a spy. It was only our great respect for military discipline that prevented a very serious demonstration at the time in his favor.

As the sad fate of young Dodd has become a part of the military history of that unfortunate struggle, it seems to me that a tribute to his memory is due from one who was then looked on as an enemy, but who recognized to the fullest the personal nobility of a character that refused to purchase life by betrayal of those who helped him procure the information found on him when arrested.

Source:

"The Trial and Execution of David O. Dodd" by Nancy Newell, an article in the Pulaski County Historical Review, Fall 1992.

A Collection of Some of
Their Stories

GUSTAV SCHURMANN

Gustav Schurmann, age 12 Regimental bugler to five generals, friend to Tad Lincoln

Gustav's story also appears in *Too Young to Die,* by Dennis Keesee. There is also a biography written by William Styple, *The Little Bugler.* Born in Westphalia, Prussia, in 1849, Gus was 11 years old when he first enlisted in the 40th New York Volunteers. Rejected at first because he was too small, his father spoke to the commanding officer, and Gustav was given a drum and told to play. The colonel said he would do. He was with the regiment from their first engagement at the Battle of Williamsburg, until they went to Harrison's Landing. There he was asked to serve General Kearny as an orderly for a day since General McClellan was to review the army.

Reporting the next day, General Kearny provided him with a horse. During the course of the review, the general had occasion to jump a large ditch. Gustav followed along, but most of the staff did not. That evening, after the review, when he reported to take his leave and return to his regiment, Gustav was told to go get his gear and bring it to headquarters and consider himself the general's orderly in the future. He also became the general's principal bugler. Kearny was killed August 31, 1862 at the Battle of Ox Hill. General David Birney replaced Kearny and kept Gustav as his bugler and orderly. Following Antietam, he was appointed to General George Stoneman's Third Corps staff and promoted to Corps bugler.

Following the Battle of Fredericksburg in December, Birney was replaced by General Daniel Sickles, who in turn promoted Gustav to sergeant, owing to his gallant service. During the grand review of the Army of the Potomac, April 9, 1863, President Lincoln noticed Gustav riding beside the general. The president's son, Tad, also noticed Gustav. The two met and an invitation to the White House followed. Gustav received an extended furlough and spent many happy days with Tad.

The Battle of Chancellorsville followed in which General Sickles was nearly cut off and captured. A daring plan, which commenced at the sound of Gustav's bugle, resulted in a successful fight to return to the main army. Following the Federal defeat and in order to rebuild spirit, a medal was created in honor of Kearney, the "Kearny Cross of Honor," and presented to five hundred select men for bravery and good conduct. Gustav was a recipient; it was a proud moment.

Nearly two months later, the army was headed to Gettysburg where General Lee's army was engaged in battle. General Sickle's corps advanced to a place in front of the Union line on the afternoon of July 2nd and came under a fierce attack near a farmhouse. The general was struck below the knee by a cannon ball and Gus placed a tourniquet on the leg. He went with the general to the hospital and then, back to Washington. President Lincoln visited them, telling of the victory at Gettysburg. The president ended Gus's military career and sent him home to his family to attend school and prepare to enter West Point. But all prospects of West Point ended with Lincoln's death. Gustav attended the funeral when the train passed through New York City, then went on with life after the war. He never saw Tad again.

Gustav Schurmann reminisces on his former general, Philip Kearny
The following passage was written by Gustav A. Schurmann, subject of the book The Little Bugler by William Styple. Schurmann served under General Kearny as an orderly and boy bugler during the Peninsular Campaign in the summer of 1862. In his adulthood, Schurmann wrote a bit about his war-time experiences. Looking back upon his days as a

soldier with the 40ᵗʰ New York Mozart Regiment, Schurmann reminisces on personal incidences relating to his association with the general:

"I will try and detail, in the smallest possible compass, as far back as I can recollect, my experience with General Kearny. In the first place, I will begin with my enlistment. In the early part of 1861, I was drumming recruits in Chatham Square, New York City, for the Forty-second Regiment Volunteers (Tammany), for a couple of months, when my father enlisted in the Fortieth N.Y. Volunteers (Mozart) at Yonkers. With the Forty-second not treating me well, I left them, not being mustered in, and tried to join the Fortieth. But its commander, Colonel Riley, would not take me on the account of my being too small and also too young, being only eleven years old. As soon as Colonel Riley said "no" I began to cry, and turned away from the tent, but my father went and spoke to him. Then he called me back and made me take a drum and a beat. All the men commenced to laugh because the drum was nearly as big as myself, but nevertheless the colonel said I would do.

"I was with the regiment from the Battle of Williamsburg, our first fight, until we went to Harrison's Landing. Corporal Brown, a clerk at General Kearny's headquarters and also a member of our regiment, came to me one day stating that General Kearny ordered him to get him a drummer from our regiment to serve as an orderly for one day, as General McClellan was to review the army the next day. I reported myself the next morning early. The general received me kindly and gave me his gray horse (Baby), one that he brought from Mexico. During the review, the general had occasion to jump a very large ditch. I jumped it with him, but a great many of the officers had to cross further up. I think my jumping this ditch brought me favorably to his notice. Accordingly, when I reported myself in the evening after the review, so as to return to my regiment, he said, "No, but go and bring your baggage over to headquarters and consider yourself my orderly in the future."

"From that day until his death I was always with the general. It was his habit to ride outside of the picket-guard every day at Harrison's Landing, only taking me with him. Many a time I would have to ride on top of the horse, lengthwise, so as not to knock my legs against the trees. He would go so fast through them, one time my hat was knocked off. As the general never stopped, by the time I was in the saddle again there was no general to be seen. But I gave "Baby" his own way and in less than five minutes he brought me up to him. I have known that same horse to kick at him as he went in the gate. The general would then "damn" me for not holding the horse tight, but for all that the general always treated me the same as my own father would have done, and no one mourned his untimely death more than I did.

Gustav Schurmann reminisces on his friendship with Tad Lincoln

"As I look back I can see that I must have been an object of envy to Tad, as by that time I had become quite a horseman, could blow a bugle, beat a drum, and swagger about like the bigger ones. The men, with whom I was somewhat of a favorite, had presented me with a mustang that had formerly been ridden by Mosby, the guerrilla chieftain, and on him I cavorted about the field until Tad could stand it no longer, and persuaded a cavalryman to lend him his horse to ride. Finally the President and Mrs. Lincoln being ready to return to Washington called Tad and bade him take leave of me."

"Mother," says Tad, I won't go home unless 'Gus' (as he already called me) can go along."

"Oh, no," interposed the president, "that won't do. This lad is a soldier, and must remain here and attend to his duties."

"I don't care, pop," responded Tad. "I want him to go home with me and teach me to ride and blow the bugle."

"This appeal, and the tears which suffused his eyes, was too much for the tender heart of our president, who ever loved Tad as the apple of his eye, and to relieve the great and good man from embarrassment General [Sickles] said: "Mr. President, if you desire, the bugler may accompany you. I will give him a furlough."

"Tad, greatly overjoyed, thanked the general, while I returned to my tent and secured my knapsack. I rode to Washington in the president's carriage, and that night slept serenely in the guests' chamber at the White House. Tad slept in a crib alongside his parents' bed. The contrast of my new quarters with my humble and sometimes uncomfortable lodgings of the past year was so overwhelming that even now the thought of the beautiful chamber that I occupied awes me.

Tad was a generous-hearted, sweet-tempered lad, with an adventurous and inventive turn of mind. I well remember one Sunday afternoon when the rain kept us in doors, that Tad's budding genius took a particularly distinctive turn, when with his little hatchet – perhaps the same one used by George Washington – he hacked at various pieces of furniture, and finally sawed away the banisters of the main stairway. When this was reported to the president, he called Tad and myself into his room and entertained us with a story about the Black Hawk war, and showed us the sword he carried in that campaign of a company of volunteers. He did not allude to our vandalism.

Tad and I owned Washington for several weeks, doing pretty much as we pleased."

Gustave Schurmann

WASHINGTON HOUSEKEEPER POTTS

Brought back from obscurity, the unidentified drummer boy of Independent Battery I has been identified.

According to his grandson, 92-year-old John Mimnall of Columbia and other local descendants, the boy is Washington Housekeeper Potts.

Born in 1848 in the Christiana area near the Chester County line, he was 14 at the time the photo was taken. A bugler as well as drummer, the boy was well known for being able to play 36 different bugle calls.

After the war, Potts went to work for Conestoga Traction Co. He died in 1911 or 1912, according to Mimnall.

Potts is buried in Riverview Cemetery.

ROBERT HENDERSHOTT,
"DRUMMER BOY OF THE RAPPAHANOCK."

Carte de Visite Portrait of Robert Hendershott, "Drummer Boy of the Rappahanock." By Kerton & Barker, New York. Hendershott wears shell jacket with standup collar and piping, kepi with star on the front, holding drum sticks in his hands. Hendershott had a remarkable history, enlisting in the 8th Michigan Inf. on three separate occasions, the third under an assumed name, and received his nickname as the result of his gallantry at Fredericksburg.
Source: HA.com auction

Thousands of Michigan lads served in the ranks, including Robert Hendershot of the 8th Michigan Infantry, who became a national celebrity by dint of a single audacious act. According to an account that first appeared in the Detroit Free Press and then was reprinted in other papers and the magazine Youth's Companion, the 12-year-old drummer from Cambridge had clung to the side of a boat as Michigan troops crossed the icy Rappahannock River at Fredericksburg, Va., on Dec. 11, 1862. Once on the other bank, Hendershot's drum was "blown to atoms," but he managed to capture one of the enemy. Hendershot reportedly had earned the praise of a general at the scene, who declared, "Boy, I glory in your spunk."

Hendershot, a fatherless hellion who found himself an object of admiration for the first time, spent the rest of the war — indeed, the rest of his life — capitalizing on his fame. He accepted an expensive and ornate drum from a New York newspaper, traveled to England to be showered with praise, appeared at P.T. Barnum's museum of curiosities, and visited the White House to meet Abraham Lincoln. He posed for countless photographs, shamelessly promoting himself as "the most wonderful Drummer in the World." Several poems were written about him, including "The Hero of the Drum." Hendershot's presence at a recruiting rally in Michigan "created much enthusiasm," the Free Press reported. More than a few wide-eyed boys in attendance could envision themselves standing on the platform, boasting of his exploits, and soaking in the admiration and the applause, just like "The Drummer Boy of the Rappahannock."

Years later, it would develop that Hendershot, by now a staple of the convention circuit, was a fraud, the tale he had weaved for newspapers wholly discounted by members of the Michigan regiments on the scene.

The real story of his time in service revealed a litany of discharges, desertions, and other unseemly behavior. "Worse than useless," was one veteran's appraisal of the boy who had demonstrated absolutely no ability to play an instrument or shoulder a weapon. If Hendershot was remembered for being at Fredericksburg at all, it was for joining in the looting that took place there — "the Forager of the Rappahannock," said another Michigan vet. But in the early stages of the war, the North was hungry for all the heroes it could get — authentic or manufactured. Hendershot, undaunted by the controversy, would keep on beating his own drum right up until his death in 1925.

* * *

War fever had gripped Jackson after the fall of Fort Sumter, and like many others Hendershot longed for the glory of battle. His widowed mother may also have hoped that military life might instill some discipline in her delinquent son. He was a frequent runaway, and his aversion to school was such that he could not even sign his own name. He claimed to be 10 that summer of 1861, but like many aspects of his life, that is in dispute, as various documents give birthdates ranging from early 1846 to 1851, and no less than four different birthplaces, from Michigan to New York City.

When he enlisted, Hendershot was a slight-framed boy, 4 ½ feet tall, with fair hair, hazel eyes and a ruddy complexion. He bore a deep scar under his right eye that he would submit as his first badge of courage. He soon dropped his implausible claim to have received that scar as the result of a severe wound at Shiloh. (At the time his regiment had been camped more than 600 miles away.) By the end of 1862, though, events at Fredericksburg would give him another, more believable opportunity for fame.

In the fall of 1861, Hendershot was a fixture in the camp of the Jackson County Rifles. There, he incessantly practiced his drum calls, an activity that caused at least one recruit to call him 'a perfect little pest.' He apparently accompanied the Rifles to Fort Wayne, outside Detroit, where the unit became Company C of the 9th Michigan Infantry. Robert claimed to have enlisted along with the others, but said that the mustering officer rejected him because of extreme youth. In any case, he boarded the train that carried the regiment south, either as a stowaway or as a servant to Captain Charles V. DeLand, the commander of Company C and editor of Jackson's American Citizen.

Robert formally enlisted in the 9th in March 1862, when the regiment moved from Kentucky to Murfreesboro, Tenn. He remained with Company C, which was posted at the Murfreesboro courthouse as provost guards. He was there on July 13 when Confederate Colonel Nathan Bedford Forrest launched a pre-dawn raid on the town. During the battle, Robert claimed that he fearlessly exposed himself to enemy fire, a claim later substantiated by several 9th Michigan soldiers.

The courage demonstrated by Hendershot and others proved useless, however. By the end of the day Forrest had captured the entire Union garrison. (See March 2002 ACW for an article on the raid.) Afterward, the enlisted men were paroled and sent to Camp Chase, Ohio. Soon after, on July 31, 1862, Robert was discharged, either because of wounds or for extreme youth, he would say. In fact, Hendershot was medically discharged because he suffered frequent and severe epileptic seizures, which had plagued him since early childhood.

Although his parole forbade him to fight against the Confederacy, in early September Hendershot appeared at a Detroit recruiting office. Because of the parole, he signed on with an alias, 'Robert Henry Henderson.' His critics would call that despicable, while others would say that it had been a common practice. Hendershot claimed he had done so at the urging of the recruiter, Lieutenant Michael Hogan.

At first there seemed little chance that Hendershot would find himself back on the battle line, for Lieutenant Hogan decided to retain him as his personal servant and aide. And so he remained for over two months, until the arrival of Chaplain George Taylor. Taylor developed a fondness for Robert and gained permission to have Hendershot placed under his care.

The two traveled south to Taylor's assigned Army of the Potomac unit, the 8th Michigan Infantry. At the Washington depot Taylor rescued Hendershot after he suffered a seizure and fell in front of a locomotive. He had another a few days later while standing at dress parade. It was then that Hendershot told Taylor of his discharge from the 9th Infantry and his use of an alias. Although Taylor kept Hendershot's confession secret, the boy began to suffer from his affliction so frequently that the acting regimental commander, Captain Ralph Ely, ordered him off duty and applied for his discharge.

It was then December, and while Hendershot awaited his discharge, the Army of the Potomac stood on the banks of the Rappahannock River opposite lightly defended Fredericksburg. There they had waited for more than three weeks for the engineers and material necessary to build pontoon bridges. The delay enabled General Robert E. Lee to move his Army of Northern Virginia into position. Thus, when the engineers arrived, Rebel sharpshooters thwarted their efforts. On December 11, the 7th Michigan Infantry volunteered to cross and drive the sharpshooters from their nests.

Hendershot had wandered to the riverbank that morning, and he tried to tag along with the regiment by climbing aboard a boat, but slipped and made the voyage across clinging to the gunwale. Newspaper accounts related stories of 'a drummer boy, only 13 years old, who volunteered and went over in the first boat' and who battled the Confederates and had his drum smashed by a shell. A correspondent for the Detroit Advertiser and Tribune wrote that the nameless boy belonged to the 8th Michigan Infantry.

Two weeks after Hendershot allegedly crossed the river, he was again discharged, for epilepsy. He was away from the regiment for more than 10 days. Right after the battle he traveled first to New York, then to Baltimore and Detroit, staking his claim to the title 'Drummer Boy of the Rappahannock.'

His first stops in Detroit were at the offices of the Advertiser and Tribune and the Free Press. Both published his 'strange and romantic' story. For several days he appeared at a local theater, where the crowds enthusiastically applauded the young hero's drum solos. Then he returned to Jackson. The editors of Jackson's newspapers, perhaps already familiar with the young man's propensity for self-promotion and exaggeration, chose not to repeat his tales.

In other parts of the country, though, many did believe his story. Among them was Horace Greeley of the New York Tribune, who summoned Hendershot to the city and presented him with a silver drum. Winfield Scott, the retired general-in-chief of the U.S. Army, was on hand for the event, as was P. T. Barnum. For the next eight weeks Hendershot performed at the showman's museum, and the youth was also rewarded with a scholarship to the Poughkeepsie Business College.

Hendershot did not remain long at the college, but did learn to write and signed his own name when he enlisted as a first-class boy aboard USS Fort Jackson, at Hampton Roads, Va., on April 1, 1864. From his naval service arose more tales of heroism with a shore party that destroyed a salt works near Fort Fisher. More likely was another story, that he fell overboard while in a seizure and a watchful shipmate saved him from drowning. And while Hendershot claimed to be discharged from the Navy on June 26, 1864, the ship's log listed him as a deserter.

The next few months were hectic, if Hendershot's tales are to be believed: He went on a grand tour of England, served as a Treasury Department page and undertook dangerous missions as a Union spy. Whatever the case, by war's end Hendershot had collected an impressive portfolio of letters from Maj. Gens. Ambrose Burnside, George Meade and others recommending him for an appointment to West Point. One notable endorsement came from President Abraham Lincoln, who wrote, 'I know of this boy, and believe he is very brave, manly and worthy.'

Hendershot claimed he was denied admission to the academy because of his wounds or his inability to pass the entrance exams. No application exists for him, however, in the academy's records. Instead, Hendershot returned to Poughkeepsie Business College for a brief time, during which he married Alice Blanchard, a fellow student. In 1867 he collaborated with a writer, William Sumner Dodge, who produced a 200-page biography, Robert Henry Hendershot; or, the Brave Drummer Boy of the Rappahannock. Around the same time, he moved to Omaha, Neb., and began working for the Union Pacific Railroad. In 1870 he applied for and received an appointment as postal clerk on the Lake Shore & Michigan Southern Railroad. From his office in Chicago, Hendershot then labored in relative obscurity for another decade.

His name once again became familiar to the public in 1881, when the Grand Army of the Republic newspaper, the National Tribune, sponsored a 'youngest soldier' contest. The first man nominated was Robert Henry

Hendershot. With unusual modesty, Hendershot did not refute the claims of many other, younger men to the title. Another five years would pass before he would emerge from the shadows.

In 1885, after his retirement, Hendershot took out his silver drum once again. Thereafter, the now self-promoted 'Major' Hendershot toured the country with his son, Cleveland, who played the fife. Although they principally performed at GAR functions and other patriotic gatherings, their tour also took them into Canada, and to the Kingdom of Hawaii, where they entertained Queen Liliuokalani.

By July 1891, the month Hendershot posted a letter to the National Tribune restating his claim to the title Drummer Boy of the Rappahannock as well as that of 'youngest soldier,' he was one of the best known veteran drummer boys. He was invited to lead the Michigan Department in the GAR parade during the organization's annual national encampment in Detroit that August.

There were certain old soldiers, however, who were not pleased by the fame and honors Hendershot enjoyed. One of them was the 7th Michigan's former drum major, Wilbur F. Dickerson. In a letter to the encampment's organizers, Dickerson pronounced Hendershot a fake and asked them to remove him from his place of honor. In other letters Dickerson asked members of the 7th Michigan to help him spearhead a campaign to discredit Hendershot.

More than 60,000 veterans paraded before 200,000 spectators to open the encampment. Hendershot marched at the head of the Michigan vets, tapping the cadence on his silver drum with sticks carved from the spear of an ancient Hawaiian warrior, a gift from Queen Liliuokalani. It was an auspicious moment for Hendershot, but disgrace would soon follow. Dickerson's efforts to discredit Hendershot began to pay off on the day following the parade at the reunion of the 7th Michigan Infantry, when Hendershot found himself the subject of an inquiry during which he was asked to tell his story and lay out his evidence. Members of the 7th Infantry who had crossed in the boats at Fredericksburg were questioned, and

Hendershot was cross-examined. In the end, the members of the regiment concluded that Hendershot's claims were false and stripped him of his title.

On August 8, the day after the 7th held its reunion, the 8th Michigan Infantry met. Its agenda also included a debate on Hendershot's claims. Hendershot, who was present at first, quickly departed when he realized the course upon which his comrades were headed. The 8th's judgment was even more severe than the 7th's: The regiment found him guilty of 'fraud, imposition, and construed forgeries,' as well as deserting his flag under fire. The members of the 8th formally drummed Hendershot out of the regiment.

But if Hendershot was not the Drummer Boy of the Rappahannock, then who was? The names of several former drummer boys were submitted for consideration. The men of the 7th Michigan Infantry tendered the names of two of their former drummer boys, John T. Spillaine and Thomas Robinson. The men of the 8th Michigan Infantry claimed the title rightfully belonged to Charles Gardner, who had died in 1864 from wounds received during the siege of Knoxville. The 31st Ohio Infantry nominated Avery Brown, who already bore the sobriquet 'Drummer Boy of the Cumberland.'

Although Brown may have legitimately challenged Hendershot's claim to the 'youngest soldier' title (he was said to have been only 9 years old when he entered the Army), his hold on the other title was seriously weakened by the fact that he had not stepped east of the Alleghenies during the war. Among the other nominees who had been present at Fredericksburg, Spillaine had the stronger case, since of them all only he was still living, and he won the title. The residents of Detroit awarded Spillaine a gold medal upon which was an embossed figure of a drummer boy and the inscription 'Drummer Boy of the Rappahannock.' Spillaine proudly wore the medal for the rest of his life.

Hendershot mounted his first appeal in the local media with a narrative of his heroic actions at Fredericksburg. His heroism had been widely reported in the press of the day, he said. He claimed that either Harper's or Frank Leslie's Illustrated Weekly had printed his image. He quoted letters and cited historian Benson Lossing's The Pictorial Field Book of the Civil

War, which included an account of Hendershot's heroics. To those who said he had deserted, he claimed that he had been wounded and sent to a hospital in Washington and then to Providence, R.I., where he continued his recovery in General Burnside's home.

Drum major Dickerson and others soon sent in rebuttals to the newspapers. Dickerson repeated his claim that Hendershot had not crossed the river, but had been 'found in a creek near camp, feigning a fit,' and that he later deserted and began 'traveling with a 10 show, telling great stories of his heroism.' A nameless correspondent wrote that Hendershot had not served in the Federal Army at all, but had spent the war as a member of band in Poughkeepsie. His fame was the result of a 'reportorial accident,' another said, 'perpetuated by well-intended but hasty acts of kindness…and by Hendershot's shrewdness at working the opportunity for all it was worth.'

Many reports of a drummer boy crossing the Rappahannock had appeared in the press immediately after the Battle of Fredericksburg, writers asserted, but Hendershot's name had not been connected with the incident until many days later. As for Lossing, one column stated that he 'was a weak judge of evidence' who had written his history 20 years after the war. Its many errors included assigning Hendershot to the 7th Michigan Infantry.

Throughout the encampment, the Reverend George Taylor had resisted those who wanted him to make a statement, lest it 'disturb the harmony of the occasion,' he claimed. Now that it was over, in a letter published in the Detroit Tribune on August 13, Taylor recounted the events of that day, and stated his 'firm conviction' that Hendershot was the Drummer Boy of the Rappahannock. That only added fuel to the fire and launched the controversy into the National Tribune.

The first volley came from Major Charles W. Bennett, historian of the 9th Michigan Infantry, Hendershot's first regiment. An examination of the evidence had convinced Bennett of the truth in Hendershot's claims, and in the Tribune he laid out his case. Captain Henry A. Ford, Grand Army editor for the Detroit Evening Journal, responded that his 'thorough inquiry into the facts of the case' had convinced him Hendershot was a fraud. Dickerson also continued his attack, while William Brewster, a drummer who had served with Hendershot, joined in and called him a camp follower, a scoundrel and a thief. Others soon joined the fray and kept the storm raging for months.

It was still raging at the GAR's 1892 meeting in Washington, D.C. There, the membership reaffirmed Spillaine's right to the title. Hendershot did not attend the encampment, the National Tribune stated, 'as it was clearly shown at the Detroit encampment that he was not entitled to this honor.'

But Hendershot was not going to let what he considered to be a 'black-hearted attack' pass without fighting back, and he called on the comrades in the first of his old regiments, the 9th Michigan Infantry, which passed a resolution supporting Hendershot's claims. He then attended the annual reunion of the 8th Michigan Infantry, the organization that had so recently drummed him out. Before he left, the regiment completely reversed its stance and passed unanimous resolutions that restored Hendershot into the 8th Michigan and supported his claim to his title.

Hendershot then rattled his way from coast to coast, from GAR posts to regimental reunions, winning back the support of veterans, one old man at a time. By the time of the national encampment of 1893, in Indianapolis, he had won his fight. There his title was reinstated to thunderous applause, after which former President Benjamin Harrison presented him with a diamond-studded, solid-gold medal inscribed 'Robert H. Hendershot, Drummer Boy of the Rappahannock, from G.A.R. and W.R.C. comrades, Indianapolis, 1893.' Soon afterward, Hendershot strengthened his claim with another biography, Camp Fire Entertainment: The True Story of R. H. Hendershot, Drummer Boy of the Rappahannock. Although Spillaine also continued to claim the title, and used it as a springboard to commander of the Michigan Department of the GAR in 1912, Hendershot apparently felt no further need to defend his title.

Was Hendershot a hero or a clever liar? How could he convince many of the great men of his day of his sincerity and worth, while the citizens of his own hometown viewed the tales of his spectacular exploits with extreme suspicion, if not outright disbelief? His first captain, Charles DeLand, made no mention of Hendershot in his History of Jackson

County, though he wrote with pride of many of Jackson's other Civil War heroes. Hendershot's absence was also conspicuous in Michigan in the War, the official history of the state's part in the Civil War.

Many newspapers did publish reports about a drummer boy crossing the Rappahannock; however, the initial reports were vague. They only told of a drummer boy, 13 years old, who belonged to either the 7th or the 8th Michigan Infantry. Although Hendershot could claim that only he fit the correspondents' description, in the weeks that followed many of these same correspondents began calling the tale a myth.

Too many high-ranking individuals endorsed Hendershot's claims to make them entirely spurious. Among his supporters was General Burnside, who only days after the battle wrote, 'He served under me faithfully…and at the battle of Fredericksburg displayed most distinguished courage.' Many who endorsed his claims, however, had not actually witnessed Hendershot's actions.

The same can be said of those who criticized Hendershot. They disbelieved his story because they had not seen him cross the river or perform his heroic deeds. In fact, they had not seen him at all. During the 1891 debate, most who crossed in the boats could not recall any drummer boy among them.

Only one man claimed to have seen Hendershot on the day in question — the Reverend George Taylor. How did he remember that day? Hendershot had frequently strayed from his camp during the previous 10 days, Taylor recalled, but December 11, 1862, was different. Stimulated by the occasion and the excitement, Robert had wandered much farther afield.

Taylor, alarmed by Hendershot's prolonged absence in the midst of battle, went to find him and came upon Hendershot coming back across a pontoon bridge. 'I met him with a bundle of clothes under his arm,' Taylor wrote. The story Hendershot then told Taylor was in many ways the same as the one the drummer would continue to tell for years. He told Taylor he had crossed the river by clinging to the stern of a boat, and that with others he had gone into deserted houses. He claimed to have set fire to a building, and that he had found a Rebel soldier hiding in a cellar 'to escape being forced back to the confederate camp.' The deserter asked Hendershot for help. 'So carrying his gun he assisted him and gave him up to our men.'

In another building, Taylor wrote, Hendershot said he found 'a beautiful clock' and started to bring it over to me.' Startled by a shell bursting nearby, Hendershot dropped it and it broke into pieces.

Taylor remembered, 'In all he told he did not seem [conscious] that he had done any very meritorious act, nor was there in his manner the least element apparent of anything that is necessary to constitute a hero.' Hendershot gave his account 'within the hearing of a number of persons, among them representatives of the press,' Taylor recounted. 'I have no doubt but that either from a misunderstanding of his statement, or designedly for the sake of making a sensation, the whole story originated.'

Thus, with the embellishment of war correspondents, a foraging expedition became a battle, a deserter became an armed adversary, and a shattered clock became a young hero's drum, burst by a shell. Soon after, Horace Greeley had summoned Hendershot to New York and presented the boy with the silver drum and honored him with his famous sobriquet.

Taylor argued that 'Boy that he was, being irresistibly borne upon the wave of fortune to the embrace of so many and such distinguished friends and to such privilege and honors, [it is not surprising that Hendershot] concluded that there must have been something in his exploits heroic and meritorious.' But, said Taylor, Hendershot was not the author of the story; it had originated with the press. Nor had he sought the title bestowed upon him by Greeley. Since 'the bearing of his title can injure no one,' Taylor continued, 'I would advise that he be allowed to depend upon it for his future success. 'In the words of our nation's most honored hero,' Taylor concluded, `Let us have Peace!' And let all the earth keep silent when I say that Robert Henry Hendershot is the genuine Drummer Boy of the Rappahannock.'

Taylor was undoubtedly right when he said that a good share of the story had been the fancy or embellishment of battlefield reporters, further embellished by Hendershot. His critics were most probably wrong when they claimed that another had performed the deeds. In their quest to shift glory, they forgot the times in which the story arose, the historical context that Taylor tried to bring out in his testimony.

'The battle was a disastrous one for our arms,' Taylor wrote, 'and while 'fire in the rear' papers were gloating over our calamity and pronouncing the war a failure, the loyal press found but little to cheer the heart of the despondent.' It was to 'divert the minds of the loyal from brooding over the disaster' that prompted Greeley, who '[seized] upon this report and sent it flying over the land in the columns of the Tribune,' Taylor wrote.

From a historic perspective, the story might be characterized as a media-created epic of heroism, inspired by an unremarkable episode at the Battle of Fredericksburg. Then, from out of the tale, Greeley plucked a boy and made of him a Northern icon. It was little more than mythology, but it served more than the sensationalist bent of the press; it also promoted patriotism and supported a cause. If propaganda is a legitimate weapon of war, then the worth of the Drummer Boy of the Rappahannock as a soldier may have equaled that of a regiment.

Hendershot was almost certainly the anonymous boy whose unremarkable deed inspired the story. He is without a doubt the boy on whom Greeley bestowed the title. It was he, then, who brought a brief moment of cheer and hope to a war-weary North in the dark days of late 1862.

Hendershot's self-promotion was harmless, and led to a legend that ultimately enriched American folklore. History would have undoubtedly buried the story and the title if Hendershot had not kept them alive. For that reason alone, he deserves to be remembered as the original Drummer Boy of the Rappahannock.

This article was written by Anthony Patrick Glesner and originally appeared in the January 2004 issue of America's Civil War magazine. For more great articles be sure to pick up your copy of America's Civil War.

David Wood
Company A, 6th Missouri Cavalry

David Wood was 10 years old when the war broke out. His father, Samuel, had moved the family from Ohio to Kansas in 1854 to help "Free Staters" keep slavery out of the territory. The elder Wood forged a reputation as the "Fighting Quaker." Growing up surrounded by hostility, it was natural for David to feel ready to march off with the troops when his father became lieutenant colonel of the 6th Missouri Cavalry in 1861. Wood's battalion was stationed at Rolla, from where it operated against Confederates in southern Missouri and Arkansas. At Rolla, David repeatedly begged his father to let him go along, but always was denied permission. One day while Colonel Wood was leading his men on a long march, miles from headquarters, he noticed a commotion at the column's rear. Turning his white stallion, he road through the ranks, all the while ignoring distractions his men contrived to divert his attention elsewhere. At last he made his way to the rear where David was found riding on a pony, surrounded by a group of admiring soldiers who were using their mounts to conceal the lad. David continued the story:

"He didn't say much to me. I guess he realized he might as well yield to the inevitable. From then on he kept me with him, and on January 1, 1862, at Rolla, I was regularly enlisted. My duties were principally that of an orderly, carrying dispatches here and there and sometimes going where grown men could not go."

In June 1862, he displayed budding business acumen: "I secured a cask of fresh water and some lemon extract and started making lemonade and selling it to the soldiers. The venture was so profitable that I made enough to buy some real lemons for a second batch. From this beginning I developed a sutler's outfit that made me in the neighborhood of $2,000 while I was in the army. Finding there was a great demand for small delicacies, I loaded up with everything I could think of that the men would buy. One of the generals from the main army loaned me an ambulance for the outfit, and soon I was handling quite a business. Among other things, I changed bills for the men, who allowed me 25 cents for changing a 5- or 10-dollar bill. This was robbery, of course, but was allowed throughout the army until Lincoln printed small bills called 'shin plasters' for change."

Too Young to Die by Dennis M. Keesee, pp. 33-34

George Washington Timmons/Stone

From an article in *Morning Star*, May 20, 2001, pg. 20

George Washington Stone's original name was George W. Timmons. Born in New Berne North Carolina on August 27, 1849, his father, Captain William Timmons was involved in the West Indies trade. His mother was an Irish Catholic and the couple had serious religious differences. The couple separated and the mother took the kids to New York, attempting to find employment there. Unfortunately things did not work out and George was found homeless on the streets by the Children's Aid Society. Soon George, age 8 at the time, and his brother Joe, along with 31 other children, were loaded on an 1857 "Orphan Train" which went west through Michigan and stopped at Albion.

George and his brother Joe were chosen and adopted by Simeon A. and Martha Stone, and their surname Timmons was changed to Stone. The Stones were elderly farmers who owned 120 acres of land. In those days, there were no child labor laws, and adopted "Orphan Train" children often worked long hours on the farm for their adoptive parents.

George was befriended by Albion dry goods merchant George N. Davis who became a lifelong friend. The Civil War began. Davis served as Captain of Company D of the First Michigan Sharpshooters regiment, and recruited George as musician/drummer on March 6, 1863. The boy was only 13 years old at the time! At the time, soldiers were required to be at least 18 years of age, but musicians were required to be at least 12.

Young boys could go if they had their parents permission. George actually ran away from home however, in order to join.

The Michigan Sharpshooters served gallantly throughout the War, much of it in Virginia. Stone was well liked by his fellow comrades, and they helped him to learn to read and write by bringing him books. George served throughout the war until his muster out in July, 1865. Despite several illnesses, he missed only one engagement and was recognized for meritorious conduct. Although assigned the duty as a drummer, Stone "was the bravest child that man ever saw. Whenever a fight began, Stone always shucked his drum and under some pretense or other shouldered his way into the front rank where he could pick up a Springfield." In one battle he fought on the front line and helped assist Colonel Charles DeLand of Jackson. Stone was also involved in the siege of Petersburg, Virginia through the last week of the war. Years later, Stone and two other comrades went back in 1899 to return the "Petersburg Grays" unit flag that the Sharpshooters had captured in 1865 as a war trophy.

In a twist of fate, one time during the War George received a needle case as part of a "care package" sent by Union women supporters in Pennsylvania. In it was a generic handwritten note by a girl to the Union soldier who would receive the case. After corresponding, this girl turned out to be George's long lost sister!

Following the War, Stone returned to Albion to recover from a war-related illness and lived with his friend and former Captain, George N. Davis. He attended the Albion Public Schools and subsequently two terms at Albion College. He then opened a local dry goods and grocery store in 1869 in partnership with C. J. Comstock. Stone married 16-year old Kittie Rice on August 10, 1870..

Following their marriage, the new couple moved to Petersburg, Virginia, George's old War battleground. He worked in the lumber business there for three years. In the years that followed, George worked in a great variety of businesses and occupations, including many positions in local governments. These took him to many places in many states, to as far west as North Dakota. Returning to Michigan, he finally settled in Lansing for the last years of his life.

George was very active in the Grand Army of the Republic as a member of Foster Post No. 45 in Lansing. He held many offices and participated in many events. Stone kept members of the group in touch with each other and helped organize various reunions.

On Armistice Day, November 11, 1921 at age 72, Stone came to Lansing from Battle Creek to march and play the drum in the parade with the Foster GAR Post fife and drum corps. After marching from the Grand Trunk railroad depot to the GAR hall, Stone looked tired. One comrade asked him if he was still going to march. "I'm going to drum until I die," he replied. The old soldiers played and marched on together, and while serenading some nearby disabled comrades, Stone dropped dead of a heart attack.

George Washington Timmons/Stone was buried in Mt. Hope Cemetery in Lansing next to his wife. He was survived by a son, Fred G. Stone of Alhambra, California.

FRED GRANT

Fred accompanied his father, General Ulysses S. Grant on scouting parties around Vicksburg. On one such occasion the party came under fire and a bullet went through Fred's coat. When his mother saw the hole in his coat, she packed up the family and went home. The following is an observation in Fred's own words

"The horrors of the battlefield were brought vividly before me. I joined a detachment which was collecting the dead for burial. Sickening at the sights, I made my way with another detachment, which was gathering the wounded, to a log house which had been appropriated for a hospital. Here the scenes were so terrible that I became faint, and making my way to a tree, sat down, the most woebegone twelve year old in America."

Fred Grant, son of then Lieutenant General Ulysses S. Grant, describing the scene at Vicksburg, Mississippi. Quoted in Murphy, Boys' War, 78

CHARLES MONELL
(New York)

Individual American Civil War Musician: Charles Monell

Charles Monell was a drummer boy with the 165th New York, 2nd Battalion, Duryee Zouaves. He served most of the Civil War. He had a CDV portrait taken in his Zouave uniform. Zouave uniforms were the inspiration for boys' Zouave suits. The portrait was once archived in a family album. Charles' home residence was not listed. He was 15 years old, I think, when the portrait was taken. He first enlisted at New York City as a Musician (September 10, 1862). The 165th, the 2nd battalion, Duryea's Zouaves, was originally recruited for a 9 months' term. Even in 1862, few people understood how long the War would last. The unit was called Duryea's Zouaves because an officer named Duryea was responsible for forming the unit -- I think Lieut.Col. Hiram Duryea,. The enlistment was subsequently changed to 3 years. Only six companies were recruited, principally from New York City and Brooklyn, and were mustered into the U. S. service between August-December 1862, for 3 years. He then mustered into "A" Co. NY 165th Infantry (November 28, 1862). The battalion left the state sailing for New Orleans (December 2, 1862). On its arrival was assigned to the 3rd brigade, 2nd (Sherman's) Division, 19th corps, Department of the Gulf. Four new companies joined the battalion in the field and were consolidated with the original six companies (1864). Charles mustered out at Charleston, SC still in the 165th Infantry (September 1, 1865). The officers were: Lieut.-Cols., Abel Smith, Jr., Governeur Carr, William R. French; Majs., Governeur Carr, Felix Angus, William W. Stephenson.

JACOB H. CULVER

Jacob Culver was born in Mercer County, Ohio, in 1845. His parents moved to the territory of Wisconsin when Jacob was only two years old. The family settled in the pine woods of Cheboygan County and engaged in the lumber business. The lad received his education in the primitive common schools of that wild land, and at the tender age of sixteen he heard the mutterings of the coming storm and his young soul was all on fire and at the first opportunity he enlisted in Company K, first Wisconsin volunteers as drummer boy, September 17, 1861, serving one year in that capacity and at the battle of Perryville, the color bearer being killed, Jacob dropped the drum and grabbed the flag and bore it aloft in triumph and continued to carry it through the service, and was mustered out with his regiment. He was in the sanguinary battles of Chaplain Hills, Chickamauga, Mission Ridge, Lookout Mountain and Chattanooga and also in the Atlanta campaign. Returning from the war he entered the Wisconsin University in 1866.

BRICE E. DAVIS

Company I, 23rd Missouri Infantry
Howard Courant, Date of publish unknown
Photos and information donated by Jody Clevenger
a descendent of Brice E. Davis.

Death of Brice E. Davis

After months of suffering Brice E. Davis died Monday, the 19th day of June 1905, of Bright's disease. Age 55 years, 2 months, 5 days.

For the last four weeks his condition has been hopeless and his death was looked for at almost any hour. His vitality has been wonderful and his suffering very great.

Drummer, Company I, 23rd Missouri Infantry entered service July 21, 1861 at the age of 11 years and was mustered out May 6, 1865. The picture was blown up from a tin type taken at Benton Barracks, St. Louis, Aug. 1862. He was taken prisoner at Shiloh and sent to Libby Prison. According to family history, they let him play around the outside of the prison because he was so young and so small.

Brice E. Davis was born in Pennsylvania. When quite young he removed with his folks to Missouri. At the age of eleven years he enlisted as a drummer boy in the 23d Missouri Infantry and was one of the very youngest enlisted soldiers of the War of the Rebellion. He was captured by the Confederates at Shiloh and was a prisoner of War for many months at Libby Prison. When he was exchanged he rejoined his regiment and served till the close of the war.

He learned the printers trade and engaged in the newspaper business at Macon, Mo., afterwards going to Kansas City, where he worked as compositor in the daily newspaper offices for many years.

In 1884 he came to Grenola, this county. He was publisher of the Grenola Chief for several years. In 1892 or '93 he came to Howard and lived here till his death. Much of the time he was a compositor in the Courant office, and was working for us when taken sick early this spring.

Mr. Davis had been twice married. He has a married daughter living in St. Louis. Nearly twenty years ago he was married to Miss Rebecca Tabor of Grenola, who with two children, Raymond and Hattie survive him.

The funeral was held at the home, Tuesday afternoon, under the auspices of the Grand Army of the Republic. Revs. Tifft and Searcy officiated in the religious services. The members of the Howard Band attended in uniform, sang the hymns and marched as escort to the remains, Mr. Davis having been a member of that organization.

At the cemetery, the G.A.R. conducted the burial with the beautiful ritual service of that order, closing with "taps" by the buglers.

The wife and children have the sympathy of this entire community in their sorrow.

Tommy Hubler
The Youngest Drummer-Boy.

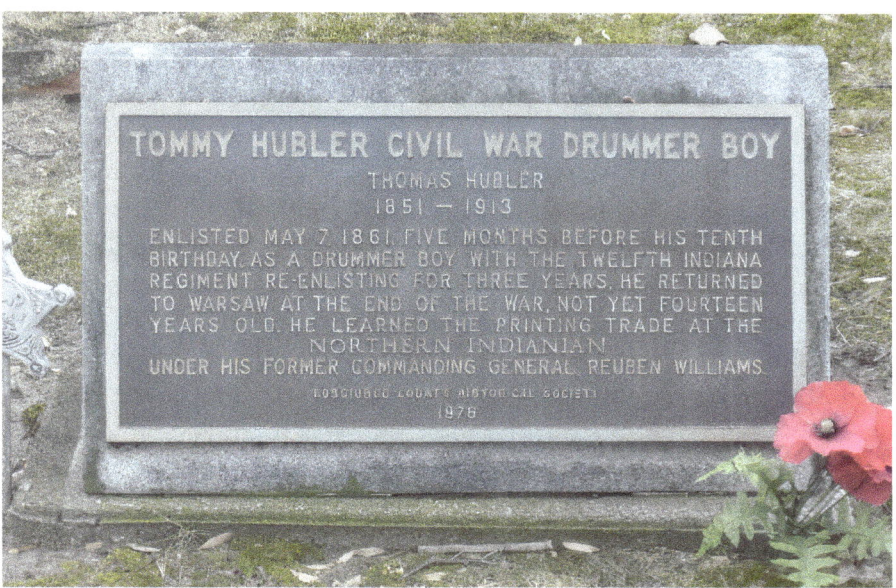

The Twelfth Indiana regiment possessed a pet of whom it may be said that he enjoyed renown scarcely second to that of the wide-famed Wisconsin eagle. This was "Little Tommy," as he was familiarly called in those days — the youngest drummer boy and, so far as the writer's knowledge goes, the youngest enlisted man in the Union army. the writer well remembers having seen him on several occasions. His diminutive size and child-like appearance, as well as his remarkable skill and grace in handling the drumsticks, never failed to fade from the memory. Some brief and honorable mention of "Little Tommy," the pride of the Twelfth Indiana regiment, should not be omitted in these "Recollections of a Drummer-boy."

Thomas Hubler was born in Fort Wayne, Allen county, Indiana, October 9, 1851. When two years of age the family removed to Warsaw, Indiana. On the outbreak of the war, his father, who had been a German soldier of the truest type, raised a company of men in response to President Lincoln's first call for 75,000 troops. "Little Tommy" was the first to enlist in his father's company, the date of enrollment being April 19, 1861. He was then nine years and six months old.

The regiment to which the company was assigned was with the Army of the Potomac throughout all its campaigns in Maryland and Virginia. At the expiration of its term of service, in August, 1862, "Little Tommy" re-enlisted and served to the end of the war, having been present in some twenty-six battles. He was greatly beloved by all the men of his regiment, with whom he was a constant favorite. It is thought that he beat the first "long roll" of the great civil war. He is still living in Warsaw, Indiana, and bids fair to be the latest survivor of the great army of which he was the youngest member. With the swift advancing years, the ranks of the soldiers of the late war are rapidly being thinned out, and those who yet remain are fast showing signs of age. "The boys in blue" are thus, as the years go by, almost imperceptibly turning into "the boys in gray;" and as "Little Tommy," the youngest of them all, sounded their first reveille, so may he yet live to beat their last tattoo. — St. Nicholas for October.

The Indiana Democrat (Indiana, Pennsylvania) Nov 15, 1883

You can read more about Tommy Hubler in this book:

Title: The Recollections of a Drummer-Boy 6th Edition (pg 160 – Google book LINK)
Author: Henry Martyn Kieffer
Publisher: Ticknor and Company, 1889

AFRICAN AMERICAN SOLDIERS IN THE CIVIL WAR: THE UNITED STATES COLORED TROOPS (USCT)

Pictured here is drummer boy Taylor of the 78th Regiment, United States Colored Infantry (USCI). While no full name is provided with the image of young Taylor, a quick examination of the National Park Service Civil

War Soldiers and Sailors System reveals there were five soldiers named Taylor who served in the 78th regiment, USCI. These Taylors were as follows: Alfred, Joseph, Nelson, Robert, and Washington. Which of these five is pictured here with his drum? That question must go unanswered for now.... In the meantime, here's a summary of young Taylor's regimental history, courtesy of the National Park Service:

78th Regiment, United States Colored Infantry

Organized April 4, 1864, from 6th Corps de Afrique Infantry. Attached to 2nd Brigade, 2nd Division, Corps de Afrique, Dept. of the Gulf, to July, 1864. Post of Port Hudson, La., Dept. of the Gulf, to October, 1864. 2nd Brigade, 2nd Division, United States Colored Troops, Dept. of the Gulf, to October, 1864. Post of Port Hudson, La., Dept. of the Gulf, to April, 1865. District of LaFourche, Dept. of the Gulf, to January, 1866.

SERVICE.--Post and garrison duty at Port Hudson, La., till April, 1865, and at Donaldsonville, Thibodeaux and other points in District of LaFourche, Dept. of the Gulf, to January, 1866. Mustered out January 6, 1866.

Predecessor unit:
CORPS DE AFRIQUE.-UNITED STATES COLORED VOLUNTEERS. 6th REGIMENT INFANTRY.

Organized at Port Hudson, La., September 4, 1863. Attached to Ullman's Brigade, Corps de Afrique, Dept. of the Gulf, to December, 1863. 2nd Brigade, 2nd Division, Corps de Afrique, to March, 1864. Garrison, Port Hudson, La., to April, 1864.

SERVICE.--Duty at Port Hudson, La., August 31, 1863. Designation of Regiment changed to 78th United States Colored Troops April 4, 1864

DAVID AULD, AGE 17

David Auld was a 17-year-old drummer boy serving alongside his brothers, Demas, 15, a drummer, and Bradford, 19. Also serving in Company B were his cousins, the Conger brothers, James, John, and Daniel.

On the second day of the battle of Corinth, October 4, 1862, the 43rd Regiment was posted to the left of Battery Robinett facing north, awaiting the charge of the dismounted 9th Texas Cavalry. As musicians, the two younger Auld brothers were given the job of stretcher bearers, when not performing drummer duties. They were to pick up the wounded and carry them to the nearby field hospital.

David Auld discusses the horrors of battle and the role the drummer boys played in caring for the wounded, dead, and dying.

"While watching these battle lines so grand to look upon, but so terrible to think of when you remember the frightful waste of human lives they caused, the call came; "Bring the stretchers, a man hurt." Myself and Demas took the stretchers to look for the man, he was pointed out to us and proved to be Bradford (our older brother) who had been struck by a shell in the left shoulder while lying on the ground in line waiting for the first assault just opening. By his side lay James W. Conger, whose clothing was stained by his blood. We were little more than children and the shock to us can be better imagined than described. Demas and myself lifted him to the stretcher just as Col. Kirby Smith and Adjutant Heyl were shot from their horses a few steps away. We carried him to the shallow ditch by the railroad a few rods to the rear, where the temporary field hospital was located, as it offered a slight protection to the wounded from the deadly hail of bullets that fell about them coming from all directions except the rear We then placed him in an ambulance still alive and conscious. We bid him goodbye and never saw him again. He only lived a short time and occupies an unknown grave,"

For the remainder of the battle the brothers worked tirelessly hauling men to the surgeon's table. When the fighting ended, they turned their attention to the Confederates on the other side of Battery Robinett.

"Of all the sights my eyes ever looked upon, this was the most ghastly and depressing. I have seen many bloody battle fields, but none have ever caused the tumult in my brain that this one did, and on no other have I ever seen in so limited a space the great numbers that were strewn and piled at Robinett and its vicinity."

*　　*　　*

told by David Auld, drummer for the 43rd Ohio Volunteers. See Civil War drummer boys did more than just play the drums. See also The History of Fuller's Ohio Brigade, 1861-1865; Its Great March, with Roster, Portraits, Battle Maps and Biographies, pg 431.

Shiloh National Military Park, March 20, 2017 at 10 AM, #visitcorinth, #Shiloh 155, #NPS

Charles William Bardeen
from his diary and letters

A civil war drummer boy cares for the wounded:
"I never want to go into a hospital again"

"When rumors of secession arose I became of course alarmed, and was always ready to express my political views to any one who would listen. One of the experiments with me was to send me up to live with a farmer named Sheldon in Peterboro, N. H., who came to Fitchburg [Massachusetts) to drive me home with him. He was so much impressed by my political harangues that he stopped one or two neighbors and set me going so that they could see what a ready tongue a boy could have. He either got tired of it or thought I was not adapted to tending sheep, for after a few days he got me into his wagon again and drove me back to Fitchburg.

"So when Sumter was fired on April 12, 1861, I was excited. I remember walking up and down the sitting room, puffing out my breast as though the responsibility rested on my poor little shoulders, shaking my fist at the south, and threatening her with dire calamities which I thought some of inflicting on her myself. I joined the military company at the Orange country grammar school and took fencing lessons. As men began to enlist I wished I were older."

Charles W. Bardeen, who was 13 in April 1861, A Little Fifer's War Diary, 17 18

". . . I was certainly scared. One shell had exploded near enough so that I could realize its effects, and the one thing I wanted was to get where no more shells could burst around me. This patriotic hero who had declared in front of campfires how he had longed for gore would have liked to be tucked up once more in his little trundle bed. Bomb ague is a real disease and I had caught it.

"There was no question of getting back to the regiment .... I could see that my division was preparing to march, and while I did not actually run I certainly walked fast to get to it. It is curious how little annoyances will keep themselves prominent even in time of danger. I had on thick woolen drawers which had somehow broken from the fastening that held them up. It was a warm day and as I hurried up the hill those drawers kept slipping down till they drove me almost distracted, disturbing my equanimity more than the danger did."

Charles W. Bardeen, a fifteen year old drummer boy with the First Massachusetts Regiment, at Fredericksburg, Virginia, in December, 1862, A Little Fifer's War Diary, 107

"Dear Mother,

"My first battle is over and I saw nearly all of it. ... Saturday the hardest fighting was done. I saw the Irish Brigade make three charges. They started with full ranks, and I saw them, in less time than it takes to write this, exposed to a galling fire of shot and shell and almost decimated.... I saw wounded men brought in by the hundred and dead men lying stark on the field, and then I saw our army retreat to the very place they started from, a loss incalculable in men, horses, cannon, small arms, knapsacks, and all the implements of war, and I am discouraged. I came out here sanguine as any one, but I have seen enough, and I am satisfied that we never can whip the South.... Let any one go into the Hospital where I was and see the scenes that I saw...."

Charles W. Bardeen, quoted in Werner, Reluctant Witnesses, 36

Here, Charles William Bardeen, drummer boy for the 1st Massachusetts Volunteers, company D, discusses his experiences during the Battle of Fredericksburg.

Bardeen writes:

"Dear Mother,

"When I closed my last it was Sunday Morning. I will relate what has passed since then. I believe I mentioned that there were several wounded Rebels brought in. As they were suffering badly, I made a Coffee pot full of coffee, giving it to all of them who wished. Most of them were in Georgia Regts, particularly the 61st & 62nd & 60th One was the Adjutant Gen'l of Erwin's Brigade, under Jackson, and in the absence of Erwin he led the Brigade in a charge upon one of our batteries. Our infantry in front united to give the batteries a chance to open with cannister, which, as soon as the enemy were near enough, they did, with terrible effect. Our infantry then advanced and took many prisoners. This Adj-Gen'l was wounded in the Groin and was in great pain. In company with all of them, he expressed great surprise at the kind treatment he received at our hands. He said he was treated as well as our own boys. All day I staid there, doing all I could for all of them. At night we went out a little way from the Hospital to sleep. I saw many legs & arms taken off, and the sight was awful. The men say that it is not battle but butchery, as the rebels are well protected by breastworks. Monday morning we were ordered back across the river, as the Div. Hospital had been established there. So the drummers were put in reliefs of six hours each to attend to the wounded. My relief is on at dark. The following were the instructions given to me by the Nurse, in the tent assigned to me. "The men on the left side will not require much attention. That man in the corner is wounded through the temple and is insane. You will have to hold him down if he attempts to get up, and you must keep close to him and keep him covered. The one next to him is crazy also. Every time he wakes up you must give him some water & look out that he does not get up. The one in this corner has got the Dysentery and will require the Bedpan often—You must pay strict attention to them all, and not let the crazy men get the upper hands of you." So off he went and left me alone with two crazy men and 6 or eight wounded ones to attend to. It was a hard place, but I did my duty as well as I was able 'till my six hours were up. I never want to go into a Hospital again."

TAD LINCOLN

This image on the left appears in an on-line search of Civil War drummer boys with text telling about young boys who served in the war. It does not identify the boy and suggests that he is unknown. However, he is known.

This photographic image from which the first image was taken, appears **in *Too Young To Die, Boy Soldiers of the Union Army*,** by Dennis M. Keesee. Mr. Keesee relates how the Secretary of War, for the fun or it, commissioned him a lieutenant. This twelve-year-old then went back to his house, dismissed the military guards, then mustered the household staff, issued them weapons, drilled them, and put them on duty. His older brother, Robert, was upset by this and went to their father. Father found it amusing and did nothing about it. The boy? Tad Lincoln. His father? The President. The author goes on to share several events. President Lincoln, his son Tad, General Grant, and his son Jesse rode out to the fort at City Point, outside Washington. Jesse's horse took off, then was chased down by Lincoln, Grant, and an orderly. After arriving, the party came under Confederate artillery fire and had to wait it out in a bombproof shelter. On another occasion, riding his pony with a cavalry boy escort, William W. Sweisfort, Tad accompanied his parents to a troop review of the 150th Pennsylvania. Sweisfort tells of his experience describing "a lively boy" who "kept

me moving." On a presidential visit at Belle Plain, Tad spied General Sickles' orderly and bugler, 12-year-old Gustav Schurmann. Borrowing a cavalryman's horse, he rode over and hung out with Gus. When the review was over, he begged that he go home with them to the White House. The president explained that he was a soldier and could not leave his command. The general stepped in and granted a furlough, releasing Gus to accompany the presidential party.

Read more about these events and others and about more boy soldiers in *Too Young To Die, Boy Soldiers of the Union Army.*

Reedsburg's Civil War Drummer Boy Buried Here

By Dorothy Douglas Parent

Frank Pettis (1850-1918) was eleven when he enlisted in the army as a drummer boy during the Civil War. At the age of twelve he began military service with his teacher, Captain A. P. Ellinwood, in the 19th Infantry, Company A. He served from February 22, 1862 to August 9, 1865.

Pettis was with his Captain in every battle in which their unit was engaged — from Suffolk, VA and Newberne, NC to the Siege of Petersburg and on to Richmond, where the colors of his regiment were the first to float from the Confederate capital building, Richmond, VA.

After the Civil War, Pettis returned to Reedsburg. First he helped in his father's tailor shop, but at the age of twenty learned the miller's trade. He was a member of the Grand Army of the Republic and the Reedsburg Drum Corps until his death on August 15, 1918. At his funeral the Reedsburg Drum Corps with muffled drums preceded the hearse to the Greenwood Cemetery where he was buried near his Captain.

Pettis left five children. One of his direct descendants, Richard Curtis Knight, lives today (1998) near Rock Springs.

ALEXANDER H. JOHNSON

Many believed that Alexander H. Johnson was the first black musician to enlist in the army during the Civil War, however, he was not really the first. Technically, at least three others have enlisted before he did.

A sailor with a passion for percussion, Johnson grew up in New Bedford, Massachusetts. For unknown reasons he was separated from his parents before his fifth birthday, and he was adopted by William Henry Johnson, the second black lawyer in the United States. Johnson's original surname was Howard and his mother was a Perry.

It is believed that William Henry Johnson's pro-military attitude most likely influenced Alexander's decision to enlist as a drummer boy in the 54th Massachusetts Infantry at the age of 16. Robert Gould Shaw, the colonel and commander of the 54th, referred to Johnson as the "original drummer boy."

Johnson was with the 54th when it left Boston for James Island, South Carolina, where it fought its first battle. Colonel Shaw fell in battle, and he was one of the 272 killed, wounded and missing out of the 600 who participated in the charge.

Johnson remained in the 54th until the end of the war. He mustered out with the survivors in the summer of 1865 and returned to Massachusetts. He brought the drum that he carried at Fort Wagner with him.

Four years later he married, settled in Worcester, Massachusetts, and organized "Johnson's Drum Corps." He led the band as drum major, and styled himself "The Major."

In 1897, a memorial to the 54th sculpted by the artist Augustus Saint-Gaudens was unveiled in Boston. The bronze relief depicts Colonel Shaw and his men leaving Boston for the South with a young drummer in the lead — a scene reminiscent of the July day in 1863 when Shaw and Johnson marched at the head of the 54th to its destiny at Fort Wagner.

After the war, Alex Johnson was a member of both the Grand Army of the Republic General George H. Ward Post #10 and of the Sons of Union Veterans of the Civil War in Worcester, Massachusetts. He is frequently mentioned in the book *We All Got History* by Nick Salvatore.

The drum carried by Johnson at Wagner remained in his possession as late as 1907, and he undoubtedly still owned it upon his death on March 19, 1930 at age 82, just a few weeks after the 67th anniversary of his enlistment in the 54th.

Written by: Henry Madison

Sources:
http://thecivilwarparlor.tumblr.com/post/76118594339/private-alexander-h-johnson-
      musician-of-the
http://battleofolustee.org/pics/alexander_johnson.html
http://opinionator.blogs.nytimes.com/2013/03/05/colonel-shaws-drummer-boy/?_r=0
http://kentakepage.com/alexander-h-johnson-the-first-drummer-boy/

GRIFFITH THOMAS

Boys, like ***Griffith Thomas*** who joined the 1st Wisconsin Heavy Artillery as a musician at 16, enlisted with dreams of adventure, but soon realized the severity of camp life — and the terrors of the battlefield. Young drummers were especially targeted by the enemy because their drumbeats communicated orders on the chaotic battlefield.

Credit: Wisconsin Veterans Museum

John January, age 14

In 1862, fourteen-year-old John January lied about his age and joined the 14th Illinois Cavalry. In July of 1864, he participated in General Stoneman's Raid, a failed attempt to attack Macon, GA, and continue south to free the prisoners at Andersonville. January, along with many others under Stoneman's command, was captured north of Macon and sent to Andersonville.

January got sick at Andersonville, but within a few months was transferred to Florence, South Carolina. At Florence, he became sick with scurvy and developed gangrene in his feet. In his biography, January describes what happened in the gangrene hospital:

"My feet and ankles, five inches above the joints presented a livid, lifeless appearance, and soon the flesh began to slough off, and the surgeon, with a brutal oath, said I would soon die. But I was determined to live and begged him to cut my feet off; telling him if he would do that I could live. He still refused and believing that my life depended on the removal of my feet, I secured an old pocketknife (I have it now in my possession) and cut through the decaying flesh, and severed the tendons.

At the close of the war, I was taken by the rebs to our lines at Wilmington, North Carolina in April 1865, and when weighed, learned that I had been reduced from 165 pounds (my weight when captured) to forty-five pounds. Every one of the Union surgeons who saw me then said I could not live; but contrary to this belief, I did, and improved."

John January recovered and later moved to South Dakota, where he lived until 1905.

Source:
An email sharing the story with all recipients
Image Credit: Library of Congress

1st Class Ship's Boy or Powder Monkey

Boys, like this one aboard the U.S.S. New Hampshire, were called powder monkeys because they ran bags of gunpowder from the stores below deck to the gun crews, moving with speed and agility. These boy assistants, as young as 10 years old, slept in hammocks below the gun decks. They were selected for their job because of their speed and height – short so they would be hidden behind the ship's gunwale, keeping them from being shot by enemy ships' sharp shooters.

George Hollat, a 16-year-old powder monkey on board the U.S.S. Varuna during an attack on Forts Jackson and St. Philip in April 1862, received the Medal of Honor for his bravery during battle. His citation reads, "He rendered gallant service through the perilous action and remained steadfast and courageous at his battle station despite extremely heavy fire and the ramming of the Varuna by the rebel ship Morgan, continuing his efforts until his ship, repeatedly holed and fatally damaged, was beached and sunk."

Twelve-year-old Henry Messhage was a 1st class boy or ship's boy or powder monkey, seen here with a bag of powder for one of the ship's guns. When not in combat, these boys served as personal assistants to the officers, cook's helpers, and general helpers for whoever needed them, assigned to whatever odd jobs needed to be done.

Credit: Massachusetts Commanders Military Order of the Loyal Legion and the U.S. Army Military History Institute, and other on line sources under "Civil War powder monkeys"

http://www.civilwarphotography.org/index.php/exhibits/online-exhibits/132-this-week-in-civil-war-photography-november-1-7

A Young Girl's Diary at the Home Front And Letters to the Home Front

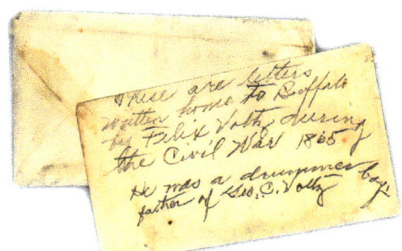

CARRIE BERRY

The following extracts are taken from the diary of Carrie Berry, a 10 year old girl from Atlanta.

We can hear the canons and muskets very plane, but the shells we dread. One has busted under the dining room which frightened us very much. One passed through the smokehouse and a piece hit the top of the house and fell through.... We stay very close to the cellar when they are shelling.

Aug, 4 The shells have been flying all day and we have stayed in the cellar. Mama put me [to work] on some stockings this morning and I will try to finish them before school commences.

Aug 5. I know all the morning. In the evening we had to run to Auntie's to get in the cellar. We did not feel safe in our cellar, they fell so thick and fast.

Aug. 6. We have been in the cellar all day....

Aug. 9. We have had to stay in the cellar all day the shells have been falling so thick around the house. Two have fallen in the garden, but none of us were hurt....

Aug. 11. Mama has ben very busy to day and I have been trying to help her all I could. We had to go to the cellar often out of the shells. How I wish the federals would quit shelling us so we could get out and get some fresh air.

Aug. 14. We had shells in abundance last night. We expected every one would come through and hurt some of us but to our joy nothing on the lot was hurt.... I dislike to stay in the cellar so close but our soldiers have to stay in ditches.

Aug. 22. I got up this morning and helped Mama pack up to move. We were glad to get our of our small cellar. We have a nice large cellar here where we can run as much as we please and enjoy it. Mama says that we make so much noise that she can't here the shells.

Aug. 23. We feel very comfortable since we have moved but Mama is fretted to death all the time for fear of fire. There is a fire in town nearly every day. I get so tired of being housed up all the time. The shells get worse and worse every day. O that something would stop them!

[September 2, 1864] Everyone has been trying to get all they could before the Federals came in the morning. They have been running with saques of meat, salt and tobacco. They did act rediculous breaking open stores and robbing them. About twelve o'clock there were a few Federals came…. In about an hour the cavalry came…. We were all frightened. We were afraid they were going to treat us badly. It was not long till the Infantry came in. They were orderly and behaved very well. I think I shall like the Yankees very well.

[Sept 10] Everyone I see seems sad. The citizens all think it is the most cruel thing to drive us from our home, but I think it would be so funny to move. Mama seems so troubled and she can't do any thing. Papa says he don't know where on earth to go.

[Nov. 16] Oh what a night we had. They came burning the store house and about night it looked like the whole town was on fire. We all set up all night. If we had not sat up our house would have been burnt for the fire was very near and the soldiers were going around setting houses on fire where they were not watched. They behaved very badly. They all left town about one o'clock this evening and we were glad when they left for nobody knows what we have suffered since they came in.

[August 1864] I was ten-years-old today. I did not have a cake. Times are too hard…. I hope that by my next birthday, we will have peace in our land.

LETTERS FROM THE CIVIL WAR:
HENRY LAWSON BERT
(AUGUST 3 & NOVEMBER 19, 1864)

Henry Lawson Bert was born at Jimstown, Ohio, on August 15, 1845, the son of Peter Bert and of Mary Frazier Bert. Henry was little more than sixteen years of age when he left his home at Tipton to enlist for the Civil War. He was not at once accepted—he was small for his age—but followed the Forty-Seventh Regiment of Indiana Volunteers from Indianapolis to Louisville, before he was finally enrolled as a drummer in Captain William M. Henley's Company I on December 21, 1861. He is described as four feet, ten inches in height, of dark complexion, with black eyes, whose occupation at the time of enrollment was that of a printer.

After the Civil War, Bert, became a merchant tailor, first in Indianapolis, and later in Edinburgh, Marion and Huntington. He died at Marion, Indiana, on December 8, 1910.

Below are the letters written by Bert.

*   *   *

Camp 47th Ind. Vet. Vol.
Morganzas Bend,
August 3rd 1864.

We are now back to this miserable hole of Morganza. I would rather stay anyplace else for we cannot get anything at All. I hope we will leave before long for I want to get away from here. The very first day we got here we got orders to go the next day on a scout and that I did not like at all but had to go. We got up at 1. o'clock and started. Marched on till after daylight and then got our breakfast and then went till 10 o'clock and then had a fight which lasted till after 12. o'clock and then we slowly retreated back with the loss of 7 men wounded out of the regiment. Eat our dinners between two and three o'clock and then went to camp. Got in about dark. Our whole march was about 32 miles. It was an awful hard day's march being very muddy coming in for it rained all afternoon. ...

*   *   *

Camp 47th Ind. Vet. Vol.,
Morganzas Bend,
August 16th, '64

Dear Sister Ann;

I now seat myself to write you a few lines this morning as it is a nice cool pleasant morning and I have not much to write but I will try to give you some of the news of camp. I guess that we will go to Mobile before long.[1] We have the promise of the first troops that leaves here. We got on a boat one night to leave and got everything loaded on and stayed on the boat till morning and then had to get off and go into camp in the same place.

We are living fine now. We get light bread every day and plenty of other rations.

We have a great deal of rain here. It has been raining now for the last three days but has cleared off now and has got to be very hot again. These coats we have got is very near wore out and the Regt. is going to send for a new suit soon again.

We have not drawed pay but once since we left home. We will have four months pay due us the last of this month and another $50.00 bounty. I dont know when we will get it but I expect not anyways soon.

Our drummers is all out drilling except me and I am on camp duty. We only have one hours drill in a day and that is at half past seven. They are out now and will be in about half past 8. o'clock and then at 9 is guard mounting and then nothing more till Dinner call and then Dress parade at six o'clock. …

Yesterday was my birthday and last night there was one of the drummers tried to pound me and he was trying to pull me down a

---

1   Admiral Farragut had defeated the Confederate fleet at Mobile on August 5, but it was not until the following March that the Forty-seventh was ordered to that point.

hill and I tore every stitch of his shirt off of him and then I got away from him and while I was gone he hid my bed close and thought I would not find them but I soon found them and went to bed.

[The Forty-seventh remained at the "miserable hole" of Morganza until September, except for an expedition to Clinton, Aug. 23 to 29. It was then sent to St. Charles, Arkansas, remaining there from Sept. 3 to Oct. 23, when it went on an expedition to Duvall's Bluff. The next move was to Little Rock, as is told in the letter of Nov. 19. The regiment did not, however, remain in Little Rock all winter— that was no more accurate than the prediction that they would be at home by the next summer. Nevertheless, the drummer-boy was not far wrong in seeing the beginning of the end in the reelection of Lincoln. He had already received the news of the re-election of Governor Morton of Indiana. On Nov. 25, only six days after this letter was written, the Regiment was returned to Memphis, remaining there until January, except for an expedition to Moscow, Dec. 21 to 31. On Jan. 1, it was ordered to New Orleans.]

*   *   *

Camp Near Little Rock, Ark.,
November 19th, 1864

Dear Sister Ann;

I now take this opportunity of trying to write you a few lines to inform you that I am still a little farther out in Arkansas near Little Rock where we have been for about three Days and I expect we will stay here till winter. When we came here we was as wet as we could be for when we got off the cars it was about nine o'clock and it looked about as much like rain as I ever saw it but it did not rain till about 1. o'clock and then it commenced raining and we was out dores [doors], had not a tent nor a house to crawl into but had to take it till about 10 o'clock the next day and then we got into some houses which some other soldiers had built and had left a few days ago and gone to fort Smith but they had left some of their sick here and they are here in the camp with us but we dont know whether we will stay in these houses all winter or not or whether we will have to go out when these other soldiers come back. There is some talk of them going some other place when they come back and if they do we will have the quarters all to ourselves but if they dont we will have to build some of our own.

Well I was speaking of the rain when we first came here and that rain is not over yet. It is still raining and has been ever since we came here. We have not seen the sun since we came here but for all the rain it does not get muddy here like it does in some places for it is very hilly and the hills is very gravely.

This is a very nice place about as nice a place as Noblesville only a great deal larger but nothing to compare with the capital of Indiana and it is very nice country around here too.

There is a great many soldiers here now and I think they will stay here all winter and; then in the Spring start out through the country after the rebels. I think that is the movements of this army but there is no telling what will be done for we have been knocked around all summer and done nothing and it may be done so this winter, but I hope we will all come home next summer and I think we will for we have won one of the grand victories already. The reason I say this is that we have got our Governor again and that is not the best yet for, we hope soon to hear of the re-election of our Noble President that we have long wished to have four years longer and when we hear of that I think we will not have to stay in the field much longer, at least I hope not.

LETTERS FROM THE CIVIL WAR:
HAMILTON WETHERBY

Born on February 13, 1847, Hamilton Wetherby was just 15 years old when he mustered into the US army on August 20, 1862, and became a drummer boy for the 111 Regiment, New York State Volunteers, Company C. Like many other young boys who seek to join the army, Hamilton lied about his age so he can serve with his brothers and cousins. Hamilton was promoted to private before he was killed in action on May 6, 1864 at the Battle of the Wilderness, Virginia. He was originally buried at Cook's farm, Spotsylvania, and then re-interred on the heights above Fredericksburg, now the national cemetery.

Below is a letter written to Hamilton's sister Ellen on Feb. 25, 1863.

*   *   *

Dear Sister Ellen,

I must begin a letter to you. I was astonished when I opened your letter and found it dated the 25 of January but you said in your letter that you did not get much time to write so I will excuse you this time but you must do better next time. You say that you are a going to school as busily as ever. It makes me almost homesick when I think of it that I cannot be there to go too. Well I will tell you about that box you sent Edwin and me. Those little pies you sent me they were dried up some but the rest of the things are first rait and those berries too were first rait. I shook them right out of the tin pail. I have that yet and I am a going to hang to it for as long as I can. Well you wanted to know if I had heard anything from Cousin David Wetherby. I have not had a letter from him since I left home. We had a paper here and it said in that that there were sixteen of David's Regt. taken prisoners and Alfred Bruicer name was called in the paper too. I am afraid that David will fair pretty hard at Richmond. I heard that they did not get only three hard crackers a day. If that is so it is pretty hard. Well, if it had not have been for General McClellan and Burnside at Harper Ferry we should have been marched to Richmond and stayed there until we were exchanged. But old Jackson had all he wanted to get away as he did.

Yes, Ellen my birth day is passed and I am sixteen years old and you are eighteen. Well I am as large as Edwin now. I should like to be there to go to singing school with you. Do you earn enough to pay you for going do you think. How much does he ask you apiece for learning you. You say Maria and Ann sends their best respects to me. Tell them I send my best respects to them and Mary too. Mothers and Emmas dress I think they are pretty. Then when school is out you must write more and oftener and I will try and do the same by you.

I must close now so good by from your Brother, Hamilton Wetherby.

Tell Father if he has got it to spar he may send me one dollar that will be enough until we get our pay. I do not want him to send it if he has not got it to spare and please let me know whether you get that check or not. Answer as soon as you get this. So good by Ellen. From Hamilton.

*   *   *

Article written by Henry Madison

Sources:

Beverly Sayles – Victory town historian. Beverly may be reached at beverlycs65@gmail.com
http://auburnpub.com/lifestyles/civil-war-drummer-boy-s-letter-gives-glimpse-into-history/
article_6a3fe384-6000-11e1-a7a2-0019bb2963f4.html
Photo Credit:
Photo taken from Wikimedia commons

LETTERS FROM THE CIVIL WAR: FELIX VOLTZ

Felix Voltz was 18 years old (possibly younger) when he ran away from home on January 30, 1865 to enlist (to his family's dismay). He mustered out with the company on July 1, 1865, at Arlington Heights, Virginia, and served as a drummer in the 187th Regiment, New York Volunteer Infantry for five months.

Felix wrote letters to his family in Elmira, New York, which describes the rigors of Union Army life from February through June 1865. The letters are held in the Special Collections Department of the University Libraries at Virginia Tech. http://spec.lib.vt.edu/

Some of the battles that Felix Voltz might have participated in were Hatcher's Run, White Oak Ridge, Five Forks and the Fall of Petersburg. He may have been at Lee's surrender at Appomattox Court House.

Below is a letter written by one of Felix's brother's to another. It recounts Felix's flight from home and includes Felix's goodbye note.

* * *

February 5, 1865

Page 1

Buffalo Feb 5th 65

Dear Brother John

In my last I told you I would tell you more but little did I think that I would have to Inform you of what has transfered this week. John it is almost beyond Expreesion what a Sad feeling prevail at the House and what a week of Sadness the week has been for our Family. Brother Felix has caused all the Sad feeling through his Headstrongness. John I will commence and give you the purticulars. last Sunday a week ago today we were all at home as usual we were all getting ready to go to church with the Exception of Brother Felix I seeing that his aim was not to go to Church so I gave him a Blowing and tryed to start him to Church but he was on unusual lenght of time dressing him so I gave him a good Blowing

Page 2

and finally he managed to get his coat on and Started off of house we all supposed he was going to church but nobody see him to church. we all came home to dinner but no Felix. we were all again to Supper but nobody had seen Felix we all began to feel uneasy and the Foalks were asking me what trouble I had with Felix and of course I told them that I tryed to start him to church of course you can Emmaggind how Mother feld and how she talked: Felix did not come home untill 10 ock when he came he went to bed. Christ asked him where he was he told him if he wanted to know he should find out. that was all that was said to him Sunday Evening when he came home. Monday morning I got up to dellifer the milk as usual and when I came home I found Felix up and Setting by the Stove. So off I gave him a Blowing for conducting himself in such a shap and putting the whole house in such an

Page 3

uprore I told him I would Brake his head if he put the house in such a shape again and in general I gave him a good Blowing instead of a Blowing he deserved a good Whipping. but he did not listen to a great many words and put on his Coat and Started for down Town. but he fetched up at the End of his Journey before night. he went and looked for some Recruting Offices and he fond them very numerius the last one of all he found at the Arcade Buildings and there the Bounty Brockers bythe name of Weaver took him to the Provost Marchalls Office where he was sworn in U.S. Service for one Years in the 187th Regt N.Y. Vols. it is the old 65th N.Y. now called the 187th. Felix left the house Monday morning and we did not know what had become of him untill the letter carrier brought us a letter Wednesday Evening from Felix I will give you the details of the letter

Page 4

Felix letter Just as he wrote it

Buffalo Jan 30 /65

Dear Parents and Sister Barbara

I thought I would write a few lines to you about my new situation I left home on Monday morning went down town and there I got in amongst some Recruting Officers and last one was in the Arcade Building is is the first one on the left hand side as you go in on Washington St. the Recruting Officers name is Weber and he enlisted me for one Year in the 187th Regh N.Y. Vols or the old 65th Regh and the place that I am in Fort Porter in Berigs one the best name is to call it Pig Pen and we are locked Just like Prisoners. and I would like to have you all come out and see me because I cant get no chance to get out and I want some little things fore use what a Soldier uses we cant get it ourselfs. the only thing we can do is sent some of the guard out and they cheat you out of half of your money No more news at present I wish you could come before I leave I expect to leave this week yet

I Remain your Truly Son

Felix Voltz

Page 5

Continued

John me and Sister went out to the Fort and found him. Thursday morning we found him in the Uniform of Unkle Sams Sister had an interview with him. he told her that he did not blame any body he says it is his own doings. he requested her to bring him something to Eat. he says he can not eat this slopp that they get to Eat. he also requested her to bring him some other nessesarys so Sister got what Articals he wanted and we went out again Friday morning and gave him the articals I see him myself Friday and tell you he regrets what he has done. Tears cam in his eyes when we left. at the presant writing Sister started for the Fort to bring him some Eatables. John Felix has attemped it 4 times but this time he has accomplished he views but he supposed he was doing us an InJury but no he is InJuring himself and not his

Page 6

Parents nor his Brothers nor his Sisters. I will tell you what has Troubled him for some time he has got sick of his Trade he was going to quit some time ago but the foalks would not let him. John you can see by his writing that he regreeted it very much already by monday when he wrote the letter. Sister has Just come Back from the Fort she has seen him he told her that he Expected to leave tomorrow for Elmira. John I gave him your address and told him to write to you if he wrtis to you you give him all the advise you ca and try to cheer him up. John there is a great many of your Rgh here at presant we see one at Church to day his name is Glever. how much pleased we would be to see you home again I will close now Respects from a great many aquantences and Love from all us at home write soon

I Remain your affectionade Brother

AW Voltz

*   *   *

Article written by Henry Madison

http://spec.lib.vt.edu/voltz/
http://thelakeshorenews.com/2013/01/18/carriage-house-stories-40/
http://chnm.gmu.edu/tah-loudoun/usregions/files/2012/08/historical-fiction.pdf
https://warriorgirl3.wordpress.com/tag/felix-voltz/

LETTERS FROM THE CIVIL WAR:
FELIX VOLTZ FEBRUARY 22, 1865

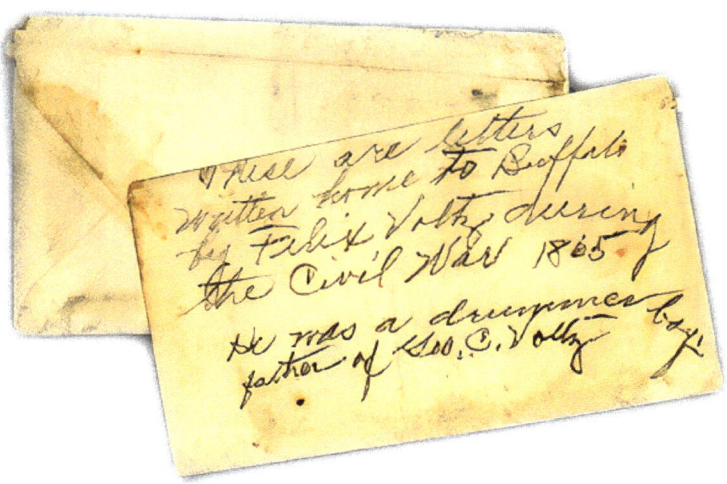

Below is a letter from Felix to his siblings describing his life in the army. More than once, Felix comments on the regret he feels for leaving home.

*   *   *

February 22, 1865

Page 1

February 22nd /65

Dear Parents Brths & Sisters

I now undertake to write you a few lines. I have so much to say that I dont what I should comence at first. I will tell you about my being here and how I got here. We left Elmira about 3ocl last Tuesday We went on the cars in Elmira and rode all that Night until the next day about 10 ocl than we came in Baltimore and we marched from the Depot up to A Fort called Federal Hill about 2 miles from the

Depot and we stayed their that Night and next day about 5 oclock then we marched all around the town untill we got to the Water and their we went

Page 2 & 3

into a Steamer but not into no Rooms or on deck for a Soldier is no Account down here at all. we was put down in lowes & end of the Boat where they keep all the dirtyest Stuff they have such things as hides and their we had to stay untill we got up to Fortess Monroe. then we got to Fortess Monroe about Noontime then we marched about 4 miles from the town and got too A place called Camp Hamilton and they put us into Barigs where there was about 300 soulders and their we stayed that Night untill Moring about 7 oclock then we marched that same way in and when we got to the dock the Boat was gone then we marced back to Fortess Monroe and stayed in the yard around untill Eveing about five oclo and then we left Fortess Monroe and rode that Night about 10 oclock then came to a place called City Point then we marched about a mile their we came too some tents and their we stayed that Night untill Next Morning about ten oclock then we went down too the Depot then we got on top the Cars and rode about 40 miles then we got to a Station called Patricks Station and their we got of and marched out in the field about 6 or 8 miles then we arrived to the Regment when we got here the whold Regment almost

Page 4-5 missing

Page 6

tell Lechlerters Boys to write to me once an while. best Respects to Stelebens Family. tell Mother not to weary about me for I am all well and Sound and I hope I will stay so this year. another I wish you let me know if you received that Picture and Photoghap and those tow Letters that I send home from Elmira and I expect the Pay Master will be here today or to morrow then I guess we will get our Bounty

I will send it as quick as I get it No more at Present only write as soon as possible for I am most sick to hear from home.

Page 7

Direct Letters
Felix Voltz Co. K
187 Regiment NYSV
2nd Division 15 Corps
Washington D.C.
To be forwarded

I remain your truly
Brother & Son
Felix Voltz

*   *   *

Article written by Henry Madison
http://spec.lib.vt.edu/voltz/
http://thelakeshorenews.com/2013/01/18/carriage-house-stories-40/
http://chnm.gmu.edu/tah-loudoun/usregions/files/2012/08/historical-fiction.pdf
https://warriorgirl3.wordpress.com/tag/felix-voltz/

Letters from the Civil War:
Felix Voltz March 3, 1865

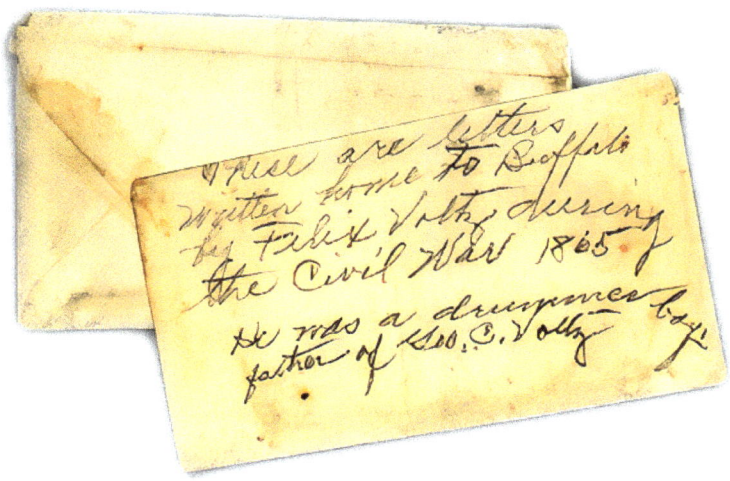

March 3, 1865

Page 1

March 3d / 65 187 Regt.
Dear Parents Brths & Sisters

I take the Pen in Hand this Evening to write you a few lines I have
so much to Say I dont know what to comence at first The first thing
I will let you know about Me being in the Drum Chor. We new
Syuat formed A new Company in this Regiment and our Officers
are for Captain is Capt. Tyler for Leit is Leit Fred Reiser for Ord is
Anthony Duncolin and I suppose you all know him the Husband of
Mary Wichter and for Drum Major is Joseph Roagh and I suppose
you all know him he used to keep a Saloon on Corner of Ellicott
& Genesee Sts.

Page 2

Now I will let you know how I got in the Drum Chor I had to go on Picket Duty the other day and when I came back I got sick for two or three days but I got over that and then I went to Tony the Orderly and ask him if they had A Drummer for our Company Says he No sir then he told me to wait A day or two and he would set about it then he to Drum Major and when he come back he told Me to go over to the Drum Major he wanted too see Me when I come over their who was Drum Major was Joe Koack and he told me if it was possible he would get Me in and then he came over and told

Page 3

me to give up my Musket and come with him then he said he would try and see if he could get Drum for Me here but he said I could not draw any government Drum down here he told Me to write Home fore one and have it send here you Tony can go and do this favor for Me he said the best and cheapest place you can buy one is on the corner of Main and Tiagarer Sts a new music Store and please buy a good one and I will make it all right as soon as I get My Bounty and he Joe told Me best way and the quickest way to send it would be by Mail / they tell us we will get our Bounty the 15th of this Month then I will send home all I possible can.

Page 4

No More news this time I will write again as soon as possible please tell Mother not to wearry herself about Me for I am all right yet and I hope will be so for the next year and tell here I am in no danger what so ever all I have to do is to take care of Me and my Drum and learn how to Drum as soon as possible I must not do no more guard or Picket Duty nor I must not take care of no Musket at all. Please tell Joe Duckene that he should excuse Me for not writing to him for I was at writing to him onced then I was called out for to go on Picket Duty but I will have More time know. So no more this time give my best Respects to all inquiring Friends. Please write as soon as possible for I am most sick to here a little something of home and please send me some postage stamps for I am out of them.

I remain your truly Son and Brother.
Felix Voltz

*   *   *

Article written by Henry Madison

http://spec.lib.vt.edu/voltz/
http://thelakeshorenews.com/2013/01/18/carriage-house-stories-40/
http://chnm.gmu.edu/tah-loudoun/usregions/files/2012/08/historical-fiction.pdf
https://warriorgirl3.wordpress.com/tag/felix-voltz/

LETTERS FROM THE CIVIL WAR:
FELIX VOLTZ MAY 20, 1865

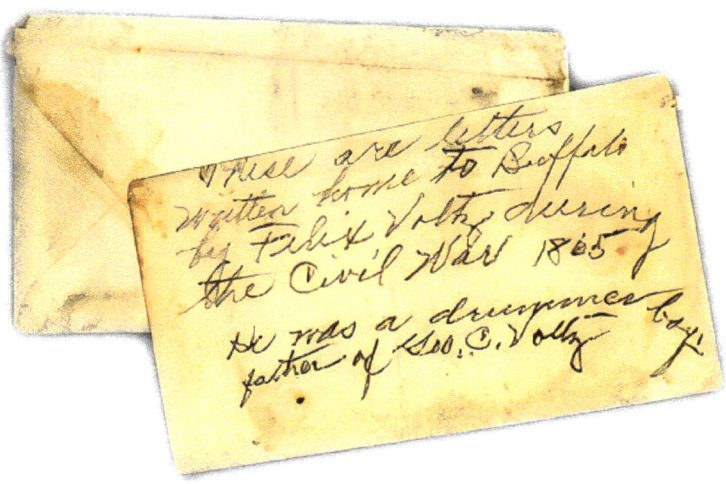

By the time Felix sent this letter, the war had ended. Troops were to be discharged—or redeployed south for Reconstruction.

*   *   *

May 20, 1865

Page 1

May 20th / 65
Camp at Arlington Hydes

Dear Brths

I take the pen in Hand to write you ounce more that I an thank God all well yet Hoping these few lines will reach you the same the reason I say ounce more is because We are all thinking of being Home in A few Days. Dr Brths I beg you to excuse me for not writing sooner. the reason was because we was on such hard Campaing and I was most all the way sick the Day we marched through Richmond it so dreadfull Hot that nobody hardly could stand it and I was sunstruck and was taken to the hospitall

Page 2

in Richmond but I got over that in two days and was send to my Regt again. Dr Brth another thing that my Drum arived here yesterday allright in good Order and Joe Roach says that you could not send A better one for here in the Army I thank you Brth A W for doing that favor and as soon I get my Pay or my Bounty I will make it all right with you. Dr Brth I wish you would Answer soon and send me some Post. St. and some paper and Envelops I know no more news at Present. I will close my writing with sending my best Regards and love to you all in the Family tell Mother not weary herself about me because I am as healthy as ever I was and tell Father that I beg him

Page 3

to forgive me for being so Ugly and Headstrong tell him that I have found out what A home is and that there is nobody on this world thank Father & Mother and A Home and tell him if God safe my Health and lets me get Home Safe again that I will try and behafe and mind my Parents better than I have. tell Lechlerters Boys that Tony Beilman has come to the Regt again he was in Hospitall since last Fall. I will close my writing by sending my best Respects to Lechlerters Boys and Family to all Duchenes Family and I wish you would let me know in what Regt Joe Duchenes is Enlisted- give my best Respects to both Uncle and Bonmans Foalks.

Page 4

So no mor news at Present
I remain your
truly Brother

Felix Voltz

*   *   *

Article written by Henry Madison

http://spec.lib.vt.edu/voltz/
http://thelakeshorenews.com/2013/01/18/carriage-house-stories-40/
http://chnm.gmu.edu/tah-loudoun/usregions/files/2012/08/historical-fiction.pdf
https://warriorgirl3.wordpress.com/tag/felix-voltz/

Letters from the Civil War:
Felix Voltz June 2, 1865

Dear Brother I have another favor to ask of you. I have the scirce on it it is what they call the Camp Edge. I ~~tried~~ tented with one of our Drummers named Fred Bernzzel and he had it back and he sent home for some Medicine he says he dont know the name of it but the Drum Major is now home on A Furlough now I supposed you have seen him before this and he had it and used some of this Medicine and it cured him and wish you would ask Joe Roach about it I wish you would see to it as soon as possible and send it by Mail if you send it by Mail it will be here in about three days I will close for this time so good day I remain your truly brother

Felix Voltz

June 2, 1865

Page 1

June 2nd / 65

Camp at Arlington Hydes

Dear Brths & Sisters

Yours dated May 24th came with great Pleasure to hand. I was very much Pleased to hear from same. I was very sorry to hear such bad news from Duchene Family Specialy to hear in what Reght Joe Enlisted in and know in his grave that their is no chance no more for me to hear from him or even to see him again. Dr Brth [Dear Brother] I must let you know that I see Brother John last week and his stayd with me all that day and that night Dear Brthr you can immagine how sursprised I was to see a Brother that I have not seen in such a long time.

Page 2

Dear Brother I have another favor to ask of you I have A leseice [lice] on Me it is what they call the Camp Edge. I tented with one of our Drummers named Fred Remmel and he had it bad and he send home for some Medicine he says he dont know the name of it but the Drum Major is now home on A furlough now I supposed you have seen him before this and he had it and used some of this medicine and it cured him and wish you would ask Joe Roach about it I wish you would see to it as soon as possible and send it by Mail if you send it by Mail it will be here in about three days I will close for this time so good day I remain your truly Brother.

Felix Voltz

*　　*　　*

Article written by Henry Madison

http://spec.lib.vt.edu/voltz/
http://thelakeshorenews.com/2013/01/18/carriage-house-stories-40/
http://chnm.gmu.edu/tah-loudoun/usregions/files/2012/08/historical-fiction.pdf
https://warriorgirl3.wordpress.com/tag/felix-voltz/

Photo taken from https://imagebase.lib.vt.edu/browse.php?folio_ID=/cw/poet

BELOW ARE JOURNAL ENTRIES AND LETTERS WRITTEN BY JOHN MCELROY AND THEODORE UPSON, DESCRIBING THEIR TIME IN THE ARMY.

Card playing had sufficed to pass the hours away at first, but our cards soon wore out…. My chum Andrews and I constructed a set of chessmen…. We found a soft white root in the swamp. A boy near us had a tolerably sharp pocket knife for the use of which a couple hours each day we gave a few spoonful of meal. The shapes that we made for pieces and pawns were crude, but sufficiently distinct for identification. We blackened one set with pitch-pine soot, found a piece of plank for a board… and so were fitted out with what served until our release to distract our attention from much of the surrounding misery.

–John McElroy, age 17, October 1863, at Andersonville, Confederate prison camp

We have been having a Christmas Jubilee. The boys raised some money and I went down into the City to get some stuff. We have a Darky cook, and he said "You alls get the greginces [ingredients] and I will get you alls up a fine dinner sure." I got some chickens, canned goods, condensed milk and a dozen eggs…. Some of the officers had a banquet – [so] they called it. I don't know if they had egg nog. If they did, their eggs must have been better than ours, but I know they must have had some sort of nog for the Provost Guard had to help some of them to their Quarters.

–Theodore Upson, 14, Indiana

LETTERS FROM THE CIVIL WAR: THOMAS VANCE (LETTERS DATED MAY 14, 1862 – SEPTEMBER 5, 1862)

Born around 1847 in Millwood, Guernsey County, Ohio, Thomas Vance enlisted in the Union Army in the 69th OVI Company I in 1862. Lying about his age to get in the army, Thomas served throughout the war and mustered-out on 19 Apr 1865 at Goldsboro, North Carolina. He was only 15 years old when he enlisted.

According to the 1850 Census he had two sisters, Mary A. and Sarah E. and two brothers, James and Asa. His parents, Thomas and Anne Vance, had a farm in Millwood.

Thomas Vance's service record entry muster out

When his parents found out that he ran away and enlisted they were horrified. They filed a suit against the local sheriff for "hijacking" their boy into uniform, however, nothing came of the suit and Thomas served his entire enlistment honorably.

After the war, in 1872 he became a Protestant Minister and in 1879 he married Melvina Elam, eventually having a family of 5 children and settling in Santa Ana, California. He was admitted to the Home for

Disabled Volunteer Soldiers in Sawtelle, California at the age of 83 on May 24, 1930 and died sometime later. He was buried in Fairhaven Cemetery, Santa Ana, California.

Nine letters that Thomas wrote during the Civil War were preserved and sent to Bowling Green State University. The letters dating between April 8, 1862 and April 3, 1864 were written to the Vance family by Thomas while he was serving with Company I, 69th Ohio Volunteer Infantry. During this period, the 69th saw action at Gallatin against Morgan (mentioned in the letter of September 5, 1862) and in such battles as Stones River at Murfreesboro, Tennessee (referred to in the letters of December 15, 1862 and January 8, 1863), Chickamauga (the 69th was primarily employed as a train guard during battle as mentioned in the letter of September 28, 1863, written at Chattanooga after the battle), and was involved in the Atlanta campaign during the period after the letters.

As you can see after reading his letters, Thomas grew and matured during his time in the army. His style of writing started out in 1862 with no capitalization and just short little "sentences". By the last letter in 1864, while his spelling and punctuation were still erratic, his comments were more astute and his style smoother.

Transcripts have been made of the letters, retaining the spelling of the originals but inserting punctuation to improve readability.

Below is Thomas' letter to his sisters dated April 8, 1862.

*   *   *

April 8th [1862]

dear sisters,

it is with pleasure that i take my pen in hand to let you know how i am a gitting along. i am well hoping these fiew lines will reach you all enjoying the same. i wrote the next day after we landed here wednesday. we are in camp in major luis about a mile from town. i expect that we will stay here to guard the town of nashvill tenisee. it is a nice country here. i like it first rate. i like souldering beter here here than i did in camp chase. it is a nicer place. the trees and grass is nice and green. there is plenty of cecesh around here and negrs. there is 50 slaves on the farm that we are on. there is a regiment of morgans cecesh some where around here. that is the report. there has ben some prisoners taken about the time that we came here. there is some around here through the woods spying. the 74 is in camp here to. i want you to write how times is and about all the news. we cant git no papers here but the nashvill papers. there is no news in them. i want you to write whither you have heard from martin vore and the rest of the souldiers and how that them is a gitting along that went to the goldmines. there was a man shot him self yesterday in the 74th. he was a blowing in it and was aplaying with the lock with his foot. i couldent get my likeness at camp chase. i will git it here if I can and send it home.

i must bring my scribling to a close for the presant. give my best respects to all inquiring friends. still remaining your true brother write soon
So good by

direct to Nashvill Tenissee in care of Captain Heslip, Co. I, 69th regiment, O.V.I.U.S.A.

Below are Thomas' letters dated May 14, 1862 – September 5, 1862. During this period, the 69th saw action at Gallatin against Morgan (mentioned in the letter of September 5, 1862).

*   *   *

Camp union, 37 miles south Nashvill
May the 14, 1862

Dear Sister

i take my pen in hand too in for you that i received your leter and was glad too hear from you and that you was well. i am well and harty and i still hope that when these few lines may com too hand they may find you all enjoying the same health. the boys is all well and in good spearits but Wells [Rufus R. Wells] and georg Mclary [George F. McClary] and they are a geting beter. We halve bin a hearing som purty good news. We heard that they was a fighting at corinth and the rebles was a vacuating it. if we Whip them thear we will halve them a bout Whiped. old morgan and his cavalry burnt a train of cars at lewisvill so we heard. we don't know wheather it is so or not. When you right too me tell me what the talk is a bout the war and what has bin a goin at hom. let me know how has the store that mr Robison had we heard that it was sole out. you wanted too know what kind of times we had. We halve the best kind of times plenty too eat and nothing too doo only too stand gard every other day and night. When you right let me know if you halve got our corn planted and what archy is a working at this spring if he is raising eney corn and how he is a getting a long with his tobaco ground. We took one prisener and captured 2 too meuls. We expect too make a good hall as soon as we com a cross eney thing that is worth taking. Thear was one compney capured 500 hundr bushel of corn a few nights a go. it was all redy sheled too send to the rebel armay. it was in a ould barn and they toock it from them. i bleave i hant very much moor too right this time onley still remain your efectionet brother untill death

Thomas Vance
Hear is a envelop all ready backed. Right soon
Sept the 5th [1862]

dear sisters

it is with pleasure that i take my seat to let you now how i am gitting
along. i am not very well nor hant been. hoping you are all stout and
hearty. i received your leter day before yesterday and was glad to hear
from home. it was wrote the 9th and 10th of aug. i hant wrote any
letters for three or four weeks. the male hant been going since the
railroad has been tour up. i reckon you herd about our galiton fight
on the lewisvill railroad. morgan took 150 prisoners and a train of
carrs with about 65 horses to. we found boxes of crackers carried
all over town by the citicens. they burnt the train. the 69th and 11th
michigan and four pieces of artillery went thare the next day and
they cidadled [skedaddled]. company A of the 69 was advance guard
they fired on some of the rebels that was out and killed three and
shot a horse. took the man prisoner. they soon left the town. we
went in town as hard as we could run. i could not keep up. we put
out pickets as soon as we got in town. they they killed two or tree.
we staid there till after noon and went back. we went on the cars
within 3 three miles of galiton. there was a bridge burnt when we
got back there in the afternoon and got on the train and was about
redy to start. the rebels came up and fierd on us killed one man. we
soon got off the cars. some of the boys on top of the cars fierd at
them. we got the artillery of the cars and fierd a fiew canister and
grape shot supposing to kill 20 or 25 not surtain. we got on the train
[the next sheet missing]

*   *   *

Article written by Henry Madison
Source:

Steve Smith (great grandson of Thomas Vance). Thomas Vance's daughter, Anna, is Steve's grandmother on his mother's side. Steve met Thomas' wife, Melvina, his great grandmother, in 1952 when she was 95 and he was five.

Steve's great uncle Ezekiel Smith also enlisted with his father (Steve's great great grandfather) during the Civil War when he was barely 16. They served in a Colorado artillery unit. Ezekiel served his entire enlistment and went on to become an Indian fighter in old west.

Additional sources:

https://lib.bgsu.edu/finding_aids/items/show/520
https://lib.bgsu.edu/finding_aids/items/show/2319

SOME OF THE BOYS WHO JOINED THE WAR WERE NOT SO LUCKY.

Below is a letter addressed to the father of a boy name Langdon Leslie Rumph, who died due to diseases at the age of 16.

My dear sir: It is with deep regret that I am compelled to inform you of the death of your son, Langdon… which occurred at the hospital yesterday morning.… He died a brave boy, and although his life was not given up in the tempest of battle, yet, he & his other deceased comrades truly deserve as much glory as those brave Southerners who fell on the bloody field of Manassas. They died in the service of their Country.… Langdon, as I presume you are aware, had been in feeble health for four or five weeks, and had just gotten over a spell of Measles when he was attacked, as his physician said, with Typhoid Fever, but I think it was a relapse from the Measles, and [he] died in five days… I have always thought that the prime causes were… the manner in which we are so crowded at this particular camp.

Benjamin Franklin Williams

Home Guard, attached to Georgia's 780th Military District
the youngest to serve the entire four years of the war in either army

From
The South's Last Boys in Gray
By Professor Jay S. Hoar
Condensed from pp. 1248-1262
used with permission

[the new Confederate marker placed May 25, 1996]
[Image taken from an episode of *Sidestep Adventures* with Robert Wright used with permission]

Benjamin Franklin Williams
March 7, 1854 – August 26, 1943

{Explanatory Note: This story has proven slow in the composing. Of some 900 aged personalities, South and North, among our nation's last surviving veterans of The War Between the States whom the writer has researched during 1971-1995, this CSA wagoner-patriot-youngster proved to be the second toughest (most challenging) of all to verify and authenticate – second only to Willie Johnston (August ?, 1850 - ? 189?), a drummer boy of the Third Vermont Infantry and early (& youngest ever) recipient of the Congressional Medal of Honor. Because the Williams Story has resurrected its own memorable odyssey-in-the-unraveling, the writer (hereafter I) would like to pass along precisely *how* its tarring truths suspensefully came to life. Hence, this present departure from an ordinary narrative mode.}

*   *   *

I believe that I first learned of Benjamin Franklin Williams' identity when in 1972 I noticed on page 279 of Pennsylvania at Gettysburg (Vol. 4, Paul L. Roy, Cmplr) "C – WILIAMS, BENJAMIN FRANKLIN, 1800 WADE AVE., N.E. Atlanta." I didn't think much about this name for quite some time, noting only that I had far fewer Ben Franklins among my 900 oldsters than I had George Washington Somebodies. The following is from a one-page featurette entitled "A Lingering Mystery Shrouds Benjamin Franklin Williams (185? – 194?) – An Active C.S.A. Patriot at Age Nine" . . . the death year pure guesswork. And it appeared in **Confederate Veteran** (March-April 1992) with this opening confession of failure.

After some nine years of fruitless research, the featurette is my ultimate effort on behalf of an apparently all-but-forgotten Confederate personality. My ambition remains the same: to discover Benjamin Franklin Williams' date and place of birth that I might fittingly honor him among the last (and youngest) grand old Boys in Gray. The one encouraging piece of evidence I have learned in recent years is in an Atlanta Constitution story (10/14/41) on the 51st National Reunion of the U.C.V. (& 46th Reunion, S.C.V.) in Atlanta. Pictured among twenty aged veterans of the Gray is "B.F. Williams, 87, of Columbus, (Ga.), fingering a golden sword that was his father's, a memento of the Mexican War; he recalls how, as just a little shaver, he drove supplies to Columbus to feed the troops there." Of Georgia's then 56 known Old Rebs, he was one of ten attending. This confirming evidence that he was indeed alive then was sent to me on April 10, 1991, by Charles Kelly Barrow, Georgia Division Historian, S.C.V., and Historian M.O.S.B., whom I wish to thank heartily.

Had it not been for an enterprising and prescient writer, Martha N. McLeod, who interviewed Cmdr. Williams, U.C.V., around 1938 for her forthcoming **Brother Warriors** *(1940), we today would most likely be unaware of Williams' humble role in the "War Between the Yankees and the Americans." What Ms. McLeod learned in 1938 directly from him we are privileged to share from her collective biography (pp. 319-20) and from mine (***The South's Last Boys in Gray***, pp. 46-47).*

"At the age of nine I was engaged in the War. My older brothers and father carried gunpowder and shots. I was too young to enlist as a soldier and carry a gun, but I did duty at the Quartermaster Supply Department throughout the four years of the war.

My father served as an enrolling officer at the beginning of the war. He got up a company of young boys and drilled them. When the last call for volunteers came, Father went to the war. The company of young boys he had prepared, gave their services by hauling supplies to the camp depots. These provisions were gathered around the country from the farmers, who gave a certain portion of their foodstuff to the quartermaster at the camp commissary. I carried lots of supplies to Wheeler's Cavalry.

One exciting come-off was when I ran a bunch of Yankees about two miles by myself. You bet, I was leading them – they tried to catch me, but I outrun 'em. They shot once in a while, just to see how fast I could run, but if they met to hit me, I outrun the bullets.

You've heard us folks down here in Georgia referred to as "Georgia Crackers;" Well, that label dates back to the ox-cart days. A long time ago traveling was mighty slow. People drove oxen hitched to little covered wagons. In order to get to town to do a little trading, a fellow had to start out in the middle of the night. Now, oxen are slow animals and sometimes just like to stop and stand still in the middle of the road. The drivers carried rawhide whips which they twirled in the air and snapped with a loud crack. The noise drove the oxen on. Lots of times the crack of the whip awakened people living along the roads to town and they would remark, 'There goes a cracker."

This then exhausts my knowledge on Cmdr. Williams. His C.S.A. service was good enough for his fellow comrades, quite obviously, and it is good enough for me. . . that I want to honor him in a forthcoming national study, "*Callow, Brave, and True*," where he surely is worthy of careful mention. It *would appear* he left no descendants. Possibly his attendant, Mrs. Langley (of Atlanta?) might have children yet who knew Cmdr. Williams? Whatever happened to him remains the haunting question. I will confer a signed hardcover on whoever is first to disclose further details on Cmdr. Williams, softbound copies upon the next two who can expand my knowledge of him and signed original tabulations from my latest unpublished study for any helpful contributor who can remove this stubborn enigma. . . .

{End of Published Query in **Confederate Veteran** magazine}

*   *   *

In response to the article a letter of April 2nd arrived from David L. Bridges (S.C.V. Camp 1347) of South Daytona, Florida, providing William's hometown of Louvale, Georgia, and his parentage, family, and burial site. Hoar then fired off a letter to the nearest newspaper to Louvale, contacting its editor, Rena Cobb. This led to a letter from one of her readers, Joseph F. (Joe) Carter, dated May 17th, who turned out to be a great grand nephew of B.F. Williams. Hoar and Carter corresponded for a year as Carter reached out through the family and local connections to gather and send the details of Ben's story along with photographs and a description of the cemetery:- *This plot lies on property they owned off west of Louvale village and fronts along the west side of a dirt road, not far from B.F.W.'s birthplace – Louvale-Holloman Creek Church Road. The graves n. to s. are B.F.W. (no d.o.d.), Nathaniel (his father), Louisa A. Cleveland (his mother), and Ida (his sister). They face west. Flat rectangular slabs cover each.*

Ben was known throughout the family as "Uncle Tobe." He married and had five children. Carter explained in the early correspondence that there was no date of death that he could find anywhere, but he was determined to find it.

A letter of June 23rd from Thelma Anna Richardson (Mrs. Thomas Lee) Wilder, a great grand niece, added more details about Uncle Tobe:-

Uncle Tobe lived near Link, "wide place in the road," and would come in to Louvale on a sled 5' long X 3' wide drawn by a mule. With a solid wood floor and low center of gravity, it had a removable straight back chair mounted mid-way. He loved children. We knew him at a distance by his big brown felt hat. . . . Uncle Tobe lived some 70 years at or near his birthplace but in the early 1930's he moved from his old home place to a small cabin he and his son Tigner had built in a hollow off the highway north of town 2.3 miles. They had a dug spring large enough to bathe in. We were baptized in it. Here tell Tigner may've had a still for his own private stock.

A lot of family history was shared by way of the two correspondents, but still no date of death.

On toward summer's end, during the month of August, Professor Hoar received an inquiry from James Gaston, Jr. of Fair Oaks Plantation, Americus, Georgia, requesting copies of his Southern study for academies with which he was affiliated. Hoar was also invited to speak at Gaston's Camp #78, S.C.V. in Americus. Plans were made for the speaking engagement to be followed by a trip to Louvale to meet with the families of Uncle Jobe. The trip took place in May of 1993.

A planned one-day visit to Louvale stretched beyond three nights, spent at the Wilder house, sleeping in an antique bed with an imposing headboard, which might well have been occupied in earlier times by Ben himself. He visited many local places, including the Williams' burial ground, and met many of the family and neighbors, some of whom had been present at Uncle Tobe's funeral.

Benjamin Franklin Williams
photo from the Professor Hoar collection
– Museum of the Bethel Historical Society

They could agree it took place in the summer of 1943, but no one could remember the exact day.

A clearer picture of what life was like for Ben during the war years emerged. He desperately wanted to be a part of his father's military home guard organization attached to Georgia's 780th Military District, which recruited boys age 9-13 with a few older men, but his parents refused. The boy persisted until they finally compromised to allow him to participate as mascot status. Presumably, he started training in mid-August of 1861, and was admitted into the unit in mid-September at the age of 7 years, 6 months, and 10 days. Officially his age was recorded as 9 years.

These boys served as mail carrier, forager, messenger, hostler, water boy, spy, lookout, produce gatherer, and wagoner. Knowing the terrain, roads, trails, waterholes, and natives made him ideal for becoming a commissary supplier in northern Stewart County and western Chattahoochee. After learning what farmers could supply what food stuffs and familiarizing himself with a penciled map which he later memorized, Ben acquired a mule and wagon cart and became a gatherer of farm produce as well as home comforts (soap, towels, clothing, etc.) and letters to the Confederate depot and outpost to an element of General Joe Wheeler's forces near the Alabama line. In warmer weather, he ran his route every ten or twelve days; one day to gather his load, another to deliver it to the Quartermaster Depot, and a third to return home. In mid-June of 1862 Ben's father relinquished his command to enlist in the infantry. For two years he stayed in the Columbus, Georgia, area and his son could visit on occasion and even spend the night. At war's end, B.F.W. was one of he youngest war veterans at 11 years and 1 month.

On the last hour at Wilder's, Hoar spoke by phone with editor Rena Cobb who shared that she was close to finding Williams' obituary and would send it on to him as soon as she had it. In her June 4, 1993, letter she wrote:- *Dear Professor Hoar: Thank you for your letter and enclosures of May 30. Luck and the Almighty were with me today. Here is B.F. William's obituary. He died in Decatur, near Atlanta, in DeKalb County, Aug. 26, 1943.*

Hoar learned later that Williams had been living for about a year in the Decatur home of his daughter, Lucile.

Professor Hoar returned to Louvale in May of 1996. On the 25th of May, 1996, a long awaited memorial service was held for Benjamin Franklin Williams and his father, Nathaniel Jackson Williams. Complete with greetings, invocation, a few words from Professor Hoar, and the placing of flowers on the graves, each with its new informative granite Confederate marker, decked over with the Confederate flag, along with the placing of a wreath of magnolia leaves by Rena Cobb. The two flags were folded and presented.

Cannons were fired in a Colorguard Salute followed by "Dixie" on guitar and a musket salute. Finally the service closed with a benediction and all returned to town for a gala community style Southern picnic.

*   *   *

NOTE: Benjamin Franklin Williams is the second youngest by five days, born March 7, 1854, five days before Robert B. Tyler, born March 12, 1854, serving as 2nd Class Boy on the ship, **USS Racer**. However, Robert didn't start his service until age 9. Upon further research in Hoar's *Our Youngest Blue and Gray*, his expanded chart of latest born lists eleven boys born even later than Robert, but only four whose service began at a younger age than B.F.W.'s. Willie H. Bush was 6 years, 5 months when he began service as valet to his father A. K. Bush at V.R.C. Elmira, New York. William Orlanda Newbold Lea began service as Cabin Boy on the **Greyhound** with the Georgia Mosquito Fleet at age 6 years, 5 months. John Hance Osteen became a Mail Carrier Aid at Santa Fe River as a member of Florida Home Guard at age 6 years, 7 months. Finally, at age 7 years, 2 months [approximately] Warren F. Dent became a Secret Service Letter Carrier in Maryland. Benjamin Franklin Williams was the only one of these youngest to serve the entire four years of the war.

This article is dedicated in memorium to Professor Jay Sherman Hoar [1933-2023] who made it his life's work to research and share the lives of the youngest to serve and the last to pass of the American Civil War. For more of Professor Hoar's research, go to the sources section.

ALBUM OF THE FORGOTTEN

[known only to a few who try to preserve their history]

[As his general watched this boy fought to stem the Federal rush-but fell, his breast pierced by a bayonet, in the trenches of Fort Mahone...Here is a boy of only fourteen years..."

Photograph from: Miller, Francis T. and Robert S. Lanier, eds. The Photographic History of the Civil War. 10 vols. New York: Review of Reviews, 1911, 3:295]

"No one's ever gone 'til they're forgotten"
Kenny Rogers, *The Gambler V, Playing for Keeps*

A collection of images of many of the boys from the war, some with names, many without, gathered from the internet, reference books, and given from private collections

young soldier

three young men

boy with great coat

young drummer

119th PA

87th PA Co D

79th Regiment

16 year old drummer

11th ri

6th infantry

1st lieutenant young Reb

boy in uniform

young soldier

young Reb

young drummer

young soldier

Confederate boy

Union boy drummer

Confederate boy [colorized]

young bugler

young drummer

young musicians

young drummer

boy drummer

young soldier

boy drummer with his sergeant

boy soldier

young soldier

young soldier

boy drummer

teen soldier

boy Zouve

young teen drummer

brothers privates Hiram & William
Gripman Co. J 3rd Minn. Inf. Reg.

Civil War drummer boy

artillery Yank

Cementous Cochran 13th
L.A. Reg. Indiana

teen soldier

Confederate teen soldier

Civil War teen drummer

drummer boy 93rd Pa

Franklin Sears

drummer boy

boy in uniform

boy on death bed

Edward I. Hager

Emzy and G.M. Taylor

Frederick G. Uthoff 16th cavalry bugler

Frank Eugene 13 YO Wisconsin

Federal drummer boy

1st class boy from Ohio

teenage German soldier

1st class boy

N.Y. Infantry teen

Nathan Jones, Camp Metcalf, Va.

Mississippi or Texas Rebel

Miles D. Shepherd, Co.
E 16th Connecticut

Marcus F. Jones, 1st
Michigan, Engineers

infantry drummer boy

Jonathan Sweet, Mississippi

teen soldier

young soldier 5th New
Hampshire infantry

boy soldier

Ohio boy

Private C.R.Battaile

Pvt. Frederick Lythson 2nd Wisconsin

Pvt. Lucien Love 43rd
Va Cavalry, Co. D

Pvt. William Hays, 16th
N.Y. Light Artillery

Rebel boy soldier

William Nelson Boswell

Jimmy Doyle Co. B 18th U.S. Infantry

boy with rifle

boy sailor

boy sailor

drummer boy

drummer boy

boys in uniform

musician tintype

drummer tintype

young soldier

Sergeant William T. Biedler
16, Co. C, Mosby's Va. Cav.

Samuel W. Doble, 12th Maine, Co. D

William R. Chessman
42nd NY Co. C & G

W. P. Ward, Co. F, 40th Ga.

Virginia Rebel

Wm. Frank Marshall age16, 19th Ind.

powder monkey [colorized]

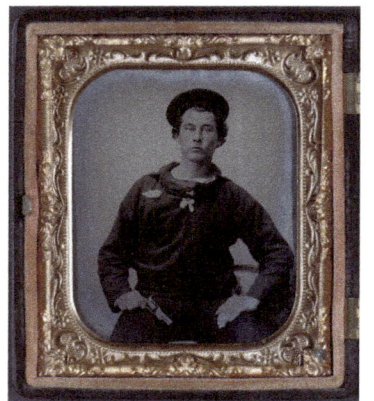

young sailor

young soldier

young soldier

young soldier

young soldier

young Confederate

young Confederate

young rebel soldier

young S. C. soldier

young Confederate soldier

young Confederate

young soldier

young soldier

Frank Eugene age 13

young soldier in Union Zouave

young soldier

young soldier

young soldier

young soldier

young Union musician

young Confederate

young drummer

Charles E. Mosby 6thVa.
Vol. Inf. Elliott Grays

young soldier

young Union soldier

young Yank

young soldier

Thomas W. Ward, 95th Pa Inf. Co. D

drummer boy

young unknown US soldier

young unknown brothers

young unknown soldier

John F. Sutton, age 15 4th
Ark Inf Mo CSA

unknown Civil War soldier

George Carleton Cassard 2nd
lieut. 10th Md. Vol. Inf.

Dennis G. Keeney 207th PA

young first class boy

37th NJ drummer

boy sailor

Christopher W. Campbell
63rd NY Co. F 15yo

drummer boy

Tennessee soldier

unidentified young soldier
in confederate uniform

William Black, drummer boy

ABOUT THE AUTHOR, J. ARTHUR MOORE

J. Arthur Moore is an educator with 42 years experience in public, private, and independent settings. He is also an amateur photographer and has illustrated his works with his own photographs. In addition to his last work, *Stranded in Snow Shoe*, Mr. Moore has written *Twelfth Winter; Journey into Darkness*, a story in four parts, *Blake's Story*, Revenge and Forgiveness, two Civil War historic fictions; and *Summer of Two Worlds*, a Native American historic fiction set in Montana Territory in the summer of 1882. *Twelfth Winter* is the sequel to *Summer of Two Worlds* and tells the story of Prairie Cub after he is forced to return to the world of his white heritage, the world of his former name, Michael. It is the emotional journey that followed. *Stranded in Snow Shoe* is the prequel, the story of the friend who set the stage for *Twelfth Winter*.

He recently published a third Civil War era historic fiction, *West to Freedom*. An earlier recent work, *Summer at Stewart Creek*, is pure fiction, set in the fictitious territory of his Virginia and Truckee Railroad of West Virginia, which he has recreated in miniature and used to illustrate this story. It is the same world in which Michael finds himself, and the beginning of the Virginia & Truckee Railroad collection of which the *Summer of Two Worlds* trilogy is a part.

The Real Boys of the Civil War is Moore's first non-fiction work. It is primarily a collection of information, assembled to provide young and old readers a glance into the world of the boys of the Civil War, to keep alive the fact that they lived, were real people, and are a part of our country's history, each serving in his own way. While some lived into their 100's, there were those, too, who never survived the war.

A graduate of Jenkintown High School, just outside of Philadelphia, Pennsylvania, Moore attended West Chester State College, currently West Chester University. Upon graduation, he joined the Navy and was stationed in Norfolk, Virginia, where he met his wife to be, a widow with four children. Once discharged from the service, he moved to Coatesville, Pennsylvania, began his teaching career, married and brought his new family to live in a 300-year-old farm house in which the children grew up and married, went their own ways, raised their families to become grandparents themselves.

Retiring after a 42-year career, Mr. Moore has moved to the farming country in Lancaster County, Pennsylvania, where he plans to enjoy the generations of family, time with his model railroad, and time to guide his writings into a new life through publication. It also allows for the opportunity to participate in a local model railroad club as well as time for traveling to Civil War events, and presenting at various organizations and events about the boys who were part of that war. He also shares the process of writing, and readings from his work, and does book signings at a variety of locations.

Mr. Moore can be reached through the contact page of the website for his books at **www.jarthurmoore.com** with links to his You Tube, Facebook and Twitter pages; and a blog page focusing on the stories of the boys who served in the Civil War.

Milton Keynes UK
Ingram Content Group UK Ltd.
UKHW022103080424
440842UK00017B/124/J